T0299894

Prospects of Hydrogen Fueled Power Generation

RIVER PUBLISHERS SERIES IN POWER

The "River Publishers Series in Power" is a series of comprehensive academic and professional books focussing on the theory and applications behind power generation and distribution. The series features content on energy engineering, systems and development of electrical power, looking specifically at current technology and applications.

The series serves to be a reference for academics, researchers, managers, engineers, and other professionals in related matters with power generation and distribution.

Topics covered in the series include, but are not limited to:

- Power generation;
- Energy services;
- Electrical power systems;
- Photovoltaics;
- Power distribution systems;
- Energy distribution engineering;
- Smart grid;
- Transmission line development.

For a list of other books in this series, visit www.riverpublishers.com

Prospects of Hydrogen Fueled Power Generation

Editors

Anoop Kumar Shukla
Department of Mechanical Engineering
Amity University Uttar Pradesh, Noida, India

Onkar Singh
Department of Mechanical Engineering
Harcourt Butler Technical University, Kanpur, India

Ali J. Chamkha
Faculty of Engineering
Kuwait College of Science and Technology, Doha, Kuwait

Meeta Sharma
Department of Mechanical Engineering
Amity University Uttar Pradesh, Noida, India

Published 2024 by River Publishers
River Publishers
Alsbjergvej 10, 9260 Gistrup, Denmark
www.riverpublishers.com

Distributed exclusively by Routledge
605 Third Avenue, New York, NY 10017, USA
4 Park Square, Milton Park, Abingdon, Oxon OX14 4RN

Prospects of Hydrogen Fueled Power Generation / by Anoop Kumar Shukla, Onkar Singh, Ali J. Chamkha, Meeta Sharma.

Routledge is an imprint of the Taylor & Francis Group, an informa business

ISBN 978-87-7004-011-2 (hardback)
ISBN 978-87-7004-063-1 (paperback)
ISBN 978-10-0382-337-7 (online)
ISBN 978-10-3265-621-2 (master ebook)

Contents

Preface xi

List of Figures xiii

List of Tables xxi

List of Contributors xxiii

List of Abbreviations xxv

1 Introduction to Hydrogen Fueled Power Generation Systems **1**
Onkar Singh, Anoop Kumar Shukla, Ali J. Chamkha, and Meeta Sharma
1.1 Introduction 1
1.2 Hydrogen Production 2
1.2.1 Pyrolysis 3
1.2.2 Gasification and reforming 4
1.2.3 Photosynthesis 5
1.2.4 Water electrolysis 6
1.3 Hydrogen in Power Generation 7
1.3.1 Co-combustion (natural gas—hydrogen mix) 9
1.3.2 Blended hydrogen fuel IC engine 9
1.4 Global Trends for Green H_2 Production 11
1.5 Challenges with Hydrogen as Fuel 12

2 Polymer Electrolyte Membrane Fuel Cell (PEMFC) Membranes **17**
Arzu Göbek and Ayşe Bayrakçeken Yurtcan
2.1 Introduction 18
2.1.1 PEMFC membranes operating at low or high temperature 19

2.2 Membrane properties 21
2.3 Solid (cationic) polymer electrolyte membranes 24
2.3.1 PEM Materials 25
2.3.1.1 Perfluoro sulfonated membranes 25
2.3.1.2 Aromatic hydrocarbon-based sulfonated membranes 30
2.3.1.3 Acid—base polymer complex membranes 33
2.3.1.4 Inorganic composite membranes 35
2.4 Conductivity Mechanism 37
2.5 Membrane Preparation 39
2.6 Characterization of Polymers and Membranes 41
2.7 Conclusions 41

3 Tri-objective Optimization of a Hydrogen-fueled Hybrid Power Generation System 53
Joy Nondy and T.K. Gogoi
3.1 Introduction 54
3.2 System Description 56
3.2.1 Modeling 57
3.2.1.1 Assumptions 57
3.2.2 Energy analysis 57
3.2.3 Exergy analysis 60
3.2.4 Economic analysis 61
3.2.5 Optimization 62
3.3 Results and Discussions 64
3.3.1 Model validation 64
3.3.2 Energy results 65
3.3.3 Exergy results 65
3.3.4 Economic results 68
3.3.5 Optimization results 69
3.4 Conclusions 75

4 Prospects for Hydrogen and Fuel Cells 81
Ayşenur Öztürk and Ayşe Bayrakçeken Yurtcan
4.1 A Brief Overview of Hydrogen Energy 82
4.2 Hydrogen Production 86
4.2.1 Thermal techniques 87
4.2.1.1 Steam reforming 87
4.2.1.2 Gasification 88

4.2.2 Electrical techniques 88
4.2.2.1 Electrolysis 88
4.2.3 Hybrid techniques 90
4.2.3.1 Photo-electrochemical method 90
4.2.4 Biological techniques 90
4.2.4.1 Dark fermentation 90
4.2.4.2 Photo fermentation 90
4.3 Hydrogen Storage 90
4.4 Fuel Cells 93
4.5 Applications of Fuel Cells 95
4.5.1 Transport applications 95
4.5.1.1 Light-duty vehicles 95
4.5.1.2 Heavy-duty vehicles 99
4.5.1.3 Marine 103
4.5.1.4 Military 104
4.5.2 Stationary applications 104
4.5.3 Mobile applications 106
4.6 Hydrogen Refueling Stations 107
4.7 Conclusions: Conspectus, Prospects, Challenges, and Roadmaps 108

5 Hydrogen Production and Bunkering from Offshore Wind Power Plants for Green and Sustainable Shipping 123
Sabri Alkan
5.1 Introduction 124
5.2 Offshore Wind Power Technology 126
5.2.1 Fixed offshore wind turbines 128
5.2.2 Floating offshore wind turbines 129
5.2.3 Offshore substations 129
5.3 Green Hydrogen Production from Offshore Wind 131
5.4 Hydrogen Power for Ships 137
5.5 Hydrogen Bunkering from Offshore Wind Power Plants . . . 140
5.6 Conclusion 141

6 Liquid Organic Hydrogen Carrier System 145
Surbhi Sharma and Khushbu Gumber
6.1 Introduction 147
6.2 Shortcomings in Formic Acid-based Fuel Cells and Advantages 149

6.3 Decomposition of Formic Acid 150
6.4 Homogeneous Catalyst for Selective HCOOH Decomposition 151
6.5 Heterogeneous Catalyst for Selective HCOOH Decomposition 152
6.6 Alternative LOHC Systems 153
6.6.1 Cycloalkanes 154
6.6.2 N-Heterocycles 156
6.7 Conclusion and Outlook 157

7 Hydrogen-added Natural Gas for Gasoline Engines 163
Nishit Bedi
7.1 Introduction 163
7.2 Enhancement in Relevant Properties of CNG with the Addition of Hydrogen 164
7.2.1 Improvement in ignition energy and quenching gap 165
7.2.2 Better flame velocity and flammability limit 165
7.2.3 Higher flame temperature 166
7.3 Hydrogen -CNG Blend as a Fuel for Spark-Ignition Engine 166
7.4 Engine Operation on Hydrogen-added Natural Gas 168
7.4.1 Method of fuel introduction 168
7.4.2 Safety feature during engine operation 169
7.4.3 Phenomenon of undesirable combustion 169
7.5 Performance and Emission Characteristics 170
7.5.1 Performance characteristics 172
7.5.1.1 Torque characteristics 172
7.5.1.2 Thermal efficiency 173
7.5.1.3 Brake-specific energy consumption (BSEC) 174
7.5.2 Emission characteristics 175
7.5.2.1 Hydrocarbon emissions 176
7.5.2.2 Nitrogen oxides emissions 177
7.6 Concluding Remarks 178

8 Performance Analysis of Hydrogen as a Fuel for Power Generation 183
Karthik Kumar, Meeta Sharma, and Anoop Kumar Shukla

8.1 Introduction 185
8.2 System Layout and Description 186
8.3 Mathematical Modeling 188
8.3.1 The proposed layout of the system and the mathematical model of the system 188
8.3.1.1 Compressor model 188
8.3.1.2 Regenerator model 189
8.3.1.3 Combustion chamber model 189
8.3.1.4 Turbine model 189
8.3.2 Exergy model 190
8.4 Results and Analysis 192
8.5 Exergy Destruction and Efficiency 201
8.6 Conclusions 203

9 Fuel Cell and Hydrogen-based Hybrid Energy Conversion Technologies **207**
Pranjal Kumar and Onkar Singh
9.1 Introduction 208
9.2 Types of Fuel Cells 210
9.3 Fuel Cell Energy Conversion Technologies 210
9.4 Hydrogen Generation with Fuel Cells 211
9.5 Requirement and Development of SOFC Integrated Cycle 212
9.6 Modeling and Analysis of Solid Oxide Fuel Cell Integrated Cycle 213
9.7 Fuel Cell and Hydrogen-based Hybrid Energy Conversion 216
9.8 Conclusions 217

10 Hydrogen-powered Transportation Vehicles **223**
Razzak Khan, Rishabh Kumar, Anuj Kumar, and Anoop Kumar Shukla
10.1 Introduction 223
10.2 Hydrogen-based Transportation Engines 225
10.3 Hydrogen Production Method 226
10.3.1 Steam-methane reforming 226
10.3.2 Gasification of coal 228
10.3.3 Electrolysis 229
10.3.4 Photo-electrolysis 230

10.3.5 Hydrogen from biomass 231
10.3.6 Hydrogen from nuclear power 232
10.4 Challenges Involving Hydrogen-integrated Vehicles 233
10.4.1 Safety of hydrogen-based vehicles 233
10.4.2 Storage of hydrogen 233
10.5 H_2-integrated Internal Combustion Engines 235
10.6 Fuel Cell Integrated Transportation Engines 236
10.7 Future Possibilities for using Hydrogen-based Transportation Systems 238

11 Hydrogen Utilization for Renewable Ammonia Production (Power-to-Ammonia) 243
Alper Can Ince, C. Ozgur Colpan, Mustafa Fazıl Serincan, and Ugur Pasaogullari
11.1 Introduction 244
11.2 Ammonia Synthesis Pathways 247
11.3 Worldwide Industry Scale PtA Projects 249
11.4 Thermodynamic Assessment of PtA System 251
11.5 Conclusions and Future Directions 256

Index 263

About the Editors 265

Preface

Hydrogen is anticipated as a key energy carrier of the future energy source as the fossil-fuel reserves turn short in supply and environment-related issues are becoming seriously concerning. The exceptionally high heating value of hydrogen and its combustion yielding water upon oxidation makes it a potential fuel to meet various energy requirements. The attractive features of hydrogen make it suitable for catering to the most of the world's energy sources that are non-fossil-based and energy services. In such a "hydrogen economy," the two complementary energy carriers hydrogen and electricity become major sources. Due to environmental concerns decarbonising the earth is a major task and countries around the world have set targets for 2050. To accomplish the target, the aim is to decarbonise the major elements, using renewable sources, mainly green hydrogen.

Green hydrogen can be an excellent source of clean energy which can tackle global climate change and poor air quality. Hydrogen-based economy features can be an opportunity for any country and a good option to decarbonize several important sectors such as transportation, shipping, the global energy market and other related industries. Although the hydrogen-based economy has many challenges associated with it and answers many questions such as the production, storage, and distribution of hydrogen. In addition to zero CO2 release, hydrogen has other important appealing physical properties such as gravimetric energy content, and a wide flammability limit than conventional fossil fuels. However, it has associated limitations such as low volumetric energy density in the gaseous phase and high cost of transportation.

This book is an aggregate effort to answer the various developments and challenges related to hydrogen in production, power generation, transportation, storage, and safety. The major field of interest related to PEMFC fuel cells, hydrogen energy and conversion methods, utilization of hydrogen for power generation and other applications is discussed by different members of the scientific community working in these areas. Variety of topics covered in respect to hydrogen as a source of energy and global push for shifting to hydrogen economy is making this compiled work much more relevant and worthwhile. The usage of clean fuel in road transportation, marine and power generation sector has been discussed in detail. Furthermore, performance investigations are also revealed in terms of energy and exergy analysis of the system. The thermodynamic optimisation for a hydrogen-fueled hybrid power generation system is also proposed herein. The diversified energy economy-based associate work is focused to support and answer green and clean fuel in the creation of a multi-sector hydrogen supply chain for numerous countries.

Editors:

Dr. Anoop Kumar Shukla
Prof. Onkar Singh
Prof. Ali J. Chamkha
Dr. Meeta Sharma

List of Figures

Figure 1.1	Biomass pyrolysis.	4
Figure 1.2	Biomass gasification for getting green hydrogen.	5
Figure 1.3	Electrolysis of water for producing hydrogen.	7
Figure 1.4	Potential of oxy hydrogen as an energy source.	8
Figure 1.5	Variations of NO emissions at different volume fractions of H_2 in a hydrogen-blended diesel engine.	11
Figure 2.1	Chemical structures of perfluorinated PEMs (Peighambardoust et al., 2010; Ogungbemi et al., 2019; Rikukawa & Sanui, 2000) (Original figure).	26
Figure 2.2	Chemical structure of Nafion (Devanathan, 2008) (Original figure).	27
Figure 2.3	Chemical structures of some of these aromatic polymers (Rikukawa & Sanui, 2000; Peighambardoust et al., 2010; Li et al., 2003; Ogungbemi et al., 2019; Lee et al., 2006) (Original figure).	31
Figure 2.4	Schema showing microstructures of Nafion (Junoh et al., 2020).	33
Figure 2.5	Chemical structure of (a–d) basic polymers and(e–f) acidic polymers (Ogungbemi et al., 2019; Peighambardoust et al., 2010; Smitha et al., 2005; Kerres et al., 1999) (Original figure).	34
Figure 2.6	Chemical structure of acid/base complex polymer electrolytes (Lee et al., 2006; Rikukawa & Sanui, 2000) (Original figure).	35
Figure 2.7	Schematic of the Grotthuss-type transport mechanism (Escorihuela et al., 2020)..	38

Figure 2.8 Schematic of vehicle-type transport mechanism (Escorihuela et al., 2020). 38

Figure 2.9 Diagram of the Grotthuss-type transport mechanism of acid complex membranes (Wong et al., 2019) (Original figure). 39

Figure 3.1 Schematic of the hybrid power generation system. 56

Figure 3.2 Bar diagram showing exergy destruction rate for each component. 67

Figure 3.3 Bar diagram showing exergy efficiency for each component. 67

Figure 3.4 Bar diagram showing capital cost rate for each component. 68

Figure 3.5 Pareto front obtained from the tri-objective optimization of the PEMFC-RORC: (a) 3D view, (b) top view, (c) side view, and (d) front view. 70

Figure 3.6 Scattered distribution plots for the active cell area (A_{cell}). 71

Figure 3.7 Scattered distribution plots for the operating temperature of PEMFC (T_{FC}). 72

Figure 3.8 Scattered distribution plots for the operating pressure of PEMFC (P_{FC}). 72

Figure 3.9 Scattered distribution plots for the turbine temperature (T_1). 73

Figure 3.10 Scattered distribution plots for the current density (i). 74

Figure 3.11 Scattered distribution plots for the membrane thickness (t_{mem}). 74

Figure 3.12 Scattered distribution plots for the condenser temperature (T_4). 75

Figure 4.1 Specific energy values of the energy sources (de Miranda, 2019). 83

Figure 4.2 National Hydrogen Strategies (2021) “Used by permission of the World Energy Council” (Council, 2021). 84

Figure 4.3 Hybridization of water electrolysis with renewable energy sources and utilization patterns of the product hydrogen (reproduced from Ref. (Yodwong et al., 2020) with permission from the MDPI). 85

Figure 4.4 Classification of hydrogen production methods (Celik & Yildiz, 2017). 86
Figure 4.5 Industrial hydrogen production from steam methane reforming (reproduced from Ref. (Lee et al., 2021) with permission from the MDPI). 87
Figure 4.6 An alkaline electrolysis cell (reproduced from Ref. (Rodriguez & Amores, 2020) with permission from the MDPI). 89
Figure 4.7 Hydrogen storage methods (reproduced from Ref. (Pal, 2021) with permission from the MDPI). . . . 91
Figure 4.8 Three main branches of fuel cell applications (reproduced from Ref. (Vinodh, 2022) with permission from the MDPI). 95
Figure 4.9 Powertrain designs of commercially available electric vehicles (reproduced from Ref. (Graber et al., 2022) with permission from the MDPI). 97
Figure 4.10 Classification of heavy-duty vehicles (reproduced from Ref. (Cunanan et al., 2021) with permission from the MDPI). 100
Figure 4.11 Hybrid propulsion of a mid-size transit bus (reproduced from Ref. (D'Ovidio et al., 2020) with permission from the MDPI). 101
Figure 4.12 Hybrid power system of heavy-duty vehicle truck (reproduced from Ref. (Liu et al., 2020) with permission from the MDPI). 102
Figure 4.13 Integrated photovoltaic-fuel cell system including photovoltaic-thermal module, battery, electrolyzer, and fuel cell (reproduced from Ref. (Ogbonnaya et al., 2021) with permission from the MDPI). . . . 105
Figure 4.14 Concept design of hydrogen refueling stations (reproduced from Ref. (Perna et al., 2022) with permission from the MDPI). 107
Figure 5.1 Overview of critical offshore wind O&M activities (GL Garrad Hassan, 2013). 127
Figure 5.2 Bottom-fixed offshore wind turbines (Jiang, 2021). 128
Figure 5.3 Floating offshore wind turbines (Jiang, 2021). . . . 130
Figure 5.4 Offshore substation and cable connections (Lerch et al., 2021). 131

Figure 5.5 Hydrogen production methods (ABS, 2020). 132

Figure 5.6 Electrolysis process and essential chemical reactions (ABS, 2020). 132

Figure 5.7 Onshore electrolysis coupled with offshore wind energy (Ibrahim et al., 2022). 134

Figure 5.8 Centralized offshore electrolysis on a floating vessel (Ibrahim et al., 2022). 134

Figure 5.9 Decentralized offshore electrolysis on a floating wind turbine substructure (Ibrahim et al., 2022). . . . 135

Figure 5.10 Decentralized offshore hydrogen production in the Dolphyn project (Caine et al., 2021; ERM, 2019) . 136

Figure 5.11 Hydrogen and hydrogen-based fuel options (Grzegorz Pawelec, 2021). 137

Figure 5.12 Fuel-cell system components onboard (BV, 2022). . 139

Figure 5.13 Flowchart and schematic of a PEMFC (DNV-GL, 2010). 140

Figure 5.14 Hydrogen bunkering from the offshore wind power plant. 141

Figure 6.1 Proposed schematic diagram for formic acid-based hydrogen carrier system. 148

Figure 6.2 Formic acid storage strategy for H_2 production via utilizing CO_2. 149

Figure 6.3 General proposed mechanism of formic acid decomposition. 151

Figure 6.4 Proposed reaction mechanism for the dehydrogenation of cyclohexane. 155

Figure 6.5 Dependency of dehydrogenation temperature of the cyclohexane on N content. 155

Figure 7.1 Percentage of hydrogen in CNG for its use in spark-ignition engine (Das and Lather, 2019; Mehra et al., 2017). 168

Figure 7.2 Schematic of the experimental test engine setup. . . 171

Figure 7.3 Engine cylinder coupled to dynamometer. 172

Figure 7.4 Engine torque characteristic vs. engine speed. . . . 173

Figure 7.5 Thermal efficiency over the range of engine speed. 174

Figure 7.6 Brake-specific energy consumption. 175

Figure 7.7 Hydrocarbon emissions vs. engine speed. 176

Figure 7.8 Nitrogen oxides vs. engine speed. 177

Figure 8.1 Layout of non-recuperative gas turbine system. . . 187
Figure 8.2 Recuperative gas turbine system layout. 187
Figure 8.3 Variation in network with pressure ratio for blend of 95% natural gas and 5% hydrogen. 193
Figure 8.4 Variation in efficiency and pressure ratio for blend of 95% natural gas and 5% hydrogen. 194
Figure 8.5 Variation of network and pressure ratio for blend of 90% natural gas and 10% hydrogen. 194
Figure 8.6 Variation in efficiency with pressure ratio for blend of 90% natural gas and 10% hydrogen. 195
Figure 8.7 Variation in network with pressure ratio for blend of 85% natural gas and 15% hydrogen. 195
Figure 8.8 Variation in efficiency and pressure ratio for blend of 85% natural gas and 15% hydrogen. 196
Figure 8.9 Variation in network with pressure ratio for blend of 80% natural gas and 20% hydrogen. 196
Figure 8.10 Variation in efficiency with pressure ratio for blend of 80% natural gas and 20% hydrogen. 197
Figure 8.11 Variation in network with pressure ratio for blend 95% natural gas, 2.5% hydrogen, and 2.5% steam. 197
Figure 8.12 Variation in efficiency with pressure ratio for blend 95% natural gas, 2.5% hydrogen, and 2.5% steam. 198
Figure 8.13 Variation in network with pressure ratio for blend 90% natural gas, 5% hydrogen, and 5% steam. . . . 198
Figure 8.14 Variation in efficiency with pressure ratio for blend 90% natural gas, 5% hydrogen, and 5% steam. . . . 198
Figure 8.15 Variation in network with pressure ratio for blend 85% natural gas, 7.5% hydrogen, and 7.5% steam. 199
Figure 8.16 Variation in efficiency with pressure ratio for blend 85% natural gas, 7.5% hydrogen, and 7.5% steam. . 199
Figure 8.17 Variation in network with pressure ratio for blend 80% natural gas, 10% hydrogen, and 10% steam. 200
Figure 8.18 Variation in efficiency and pressure ratio for blend 80% natural gas, 10% hydrogen, and 10% steam. 200

Figure 8.19 Variation in exergy destruction in all the components with fuel composition. 201
Figure 8.20 Variation in exergy destruction in all the components with fuel composition. 202
Figure 8.21 Exergetic efficiencies of the components with varying fuel composition. 202
Figure 8.22 Exergetic efficiencies of the components with varying fuel composition 203
Figure 9.1 Schematic of hydrogen gas generation. 209
Figure 9.2 Representation of the voltage variation with the current density (Larminie and Dicks, 2003). 216
Figure 10.1 Multiple production processes and energy sources are connected to a wide range of fuel cell applications using hydrogen (Edwards PP, 2008). 227
Figure 10.2 Steam reformation method for producing hydrogen gas (Image courtesy: https://www.mvsengg.com/blog/steam-methane-reformer/ accessed on 27/03/2023). 228
Figure 10.3 Coal gasification method (Image courtesy: http://butane.chem.uiuc.edu/pshapley/environmental/l5/1.html/accessedon27/01/2023). 228
Figure 10.4 Process of electrolysis (Amores, Sánchez, Rojas, & Sánchez-Molina, 2021). 229
Figure 10.5 Generation of hydrogen from photo electrolysis (Image courtesy: https://bhuang02.tripod.com/photoelectrolysis.htm/accessedon27/01/2023). 230
Figure 10.6 Production of hydrogen gas from biomass gasification (Besha, 2020). 231
Figure 10.7 Hydrogen generation from nuclear reactor (Belghit*, 2020). 232
Figure 10.8 Flowchart showing hydrogen storage methods. . . . 234
Figure 10.9 (a) Pressure booster IC engine. (b) Liquid H_2 IC engine (Gurz, 2017). 235
Figure 10.10 Different types of fuel cells (Hames, 2018). 238
Figure 11.1 End-use products of ammonia. 245
Figure 11.2 A basic configuration of the PtA system process. . . 247
Figure 11.3 On average at the best sites in the world cost of ammonia synthesis through the PtA process (data reproduced based on ref. [7]). 248

Figure 11.4 A schematic configuration of the method considered in this study. 254

Figure 11.5 The change of reactor outlet temperature versus reactor inlet temperature under various operating pressures. 254

Figure 11.6 The effect of pressure on the reactor outlet compositions: (a) 10 MPa, (b) 20 MPa, and (c) 50 MPa. 255

Figure 11.7 The flow rates of ammonia synthesized and nitrogen consumed during the reaction under the operation of SOE and PEM electrolyzer for pressures of (a) 10 MPa, (b) 20 MPa, and (c) 50 MPa. 256

List of Tables

Table 1.1 Methods of hydrogen generation. 3
Table 1.2 Lower heating value and flame speed of fuels. 12
Table 2.1 Physical properties and performance of some polymer types (references in the table are valid). 23
Table 2.2 Features of commercial perfluorinated PEMs (Kim et al., 2015; Li et al., 2003; Lee et al., 2006). 28
Table 2.3 Advantages and disadvantages of the different modifications on different PEMs (references in the table are valid). 36
Table 3.1 The input parameters of the RORC (Nondy & Gogoi, 2021b). 57
Table 3.2 Energy and exergy equations used for modeling PEMFC-RORC. 60
Table 3.3 The cost equations applied to evaluate PEC_k. 62
Table 3.4 Range of parameters. 63
Table 3.5 The input variables used for running the PESA-II. . . 64
Table 3.6 Model validation for PEMFC using the data reported in Ref. (Zhao et al., 2012). 64
Table 3.7 Model validation for RORC using the data reported in Ref. (Safarian & Aramoun, 2015). 65
Table 3.8 Comparative results of PEMFC-RORC with PEMFC-ORC. 65
Table 3.9 State properties at various state points of the hybrid system. 66
Table 3.10 Energy results of the hybrid system at the base case condition. 66
Table 3.11 Exergy results of the hybrid system at the base case condition. 68
Table 3.12 The value decision variables at the best optimal solution. 69

Table 3.13 The values objective functions at the best optimal solution. 71
Table 4.1 Characteristics of basic fuel cells (reproduced from Ref. (Vaghari, 2013) with permission from the Springer). 93
Table 4.2 The properties of liquid fuels in direct liquid fuel cells (reproduced from Ref. (Öztürk et al., 2020) with permission from the Elsevier). 94
Table 4.3 Fuel cell electric vehicles status in Japan, Europe, and USA (reproduced from Ref. (Asif & Schmidt, 2021) with permission from the MDPI). 98
Table 6.1 Preferred heterogeneous catalysts for the decomposition of formic acid. 153
Table 7.1 Properties of hydrogen, methane, and gasoline (White et al., 2006; Saravanan et al., 2007; Eichlseder and Klell, 2010). 165
Table 7.2 Fuel injection methods (Das 1990). 169
Table 7.3 Specifications of single-cylinder SI engine. 171
Table 8.1 Standard molar chemical exergy values. 191
Table 10.1 Comparison of recent developed engines with fossil fuel-based engines (Butler, An overview of development and challenges in hydrogen powered vehicles, 2019). 225
Table 10.2 Hydrogen-based vehicles available globally. 237
Table 11.1 Currently operating, under construction, and planned industry-scale PtA projects. 250
Table 11.2 The sink and source terms used in mass and energy balance for the thermodynamic model. 252
Table 11.3 Reaction kinetics parameters. 252

List of Contributors

Alkan, Sabri, *Department of Motor Vehicles and Transportation Technologies, Maritime Vocational School of Higher Education, Bandırma Onyedi Eylül University, Turkey*

Bedi, Nishit, *Government Engineering College, Ujjain, India*

Chamkha, Ali J., *Dean of Engineering, Kuwait College of Science and Technology, Qatar*

Colpan, C. Ozgur, *Dokuz Eylul University, Faculty of Engineering, Mechanical Engineering Department, Izmir*

Göbek, Arzu, *Faculty of Science, Department of Chemistry, Atatürk University, Turkey*

Gogoi, T.K., *Department of Mechanical Engineering, Tezpur University, India*

Gumber, Khushbu, *Chemistry Department, University Institute of Sciences, Chandigarh University, India*

Ince, Alper Can, *Center for Clean Energy Engineering, University of Connecticut, United States; Department of Mechanical Engineering, University of Connecticut, United States*

Khan, Razzak, *School of Mechanical Engineering (SMEC), Vellore Institute of Technology, India*

Kumar, Anuj, *School of Mechanical Engineering (SMEC), Vellore Institute of Technology, India*

Kumar, Karthik, *Department of Mechanical Engineering, Amity University, India*

Kumar, Pranjal, *Harcourt Butler Technical University, India*

Kumar, Rishabh, *School of Mechanical Engineering (SMEC), Vellore Institute of Technology, India*

Nondy, Joy, *Department of Mechanical Engineering, Tezpur University, India*

Öztürk, Ayşenur, *Faculty of Engineering, Department of Chemical Engineering, Atatürk University, Turkey*

Pasaogullari, Ugur, *Center for Clean Energy Engineering, University of Connecticut, United States*; *Department of Mechanical Engineering, University of Connecticut, United States*

Serincan, Mustafa Fazıl, *Gebze Technical University, Faculty of Engineering, Mechanical Engineering Department, Turkey*

Sharma, Meeta, *Department of Mechanical Engineering, Amity University, India*

Sharma, Surbhi, *Chemistry Department, University Institute of Sciences, Chandigarh University, India*

Shukla, Anoop Kumar, *Department of Mechanical Engineering, Amity University, India*

Singh, Onkar, *Harcourt Butler Technical University, India*; *VMSB Uttarakhand Technical University, India*

Yurtcan, Ayşe Bayrakçeken, *Faculty of Engineering, Department of Chemical Engineering, Atatürk University, Turkey; Graduate School of Science, Department of Nanoscience and Nanoengineering, Atatürk University, Turkey*

List of Abbreviations

3E	Energy, exergy, and economic
AB	Ammonia borane
ABS	American Bureau of Shipping
AFC	Alkaline fuel cell
AMFC	Alkaline membrane fuel cell
APU	Auxiliary power unit
AWE	Alkaline water electrolysis
BEV	Battery electric vehicle
BSEC	Brake-specific energy consumption
BTE	Brake thermal efficiency
BV	Bureau Veritas
CCGT	Combined cycle gas turbine
CDC	Color-less distributed combustion
CHP	Combined heat and pump
CI	Continuous carburation
CL	Catalyst layer
CMI	Continuous manifold injection
CNF	Carbon nanofibers
CNG	Compressed natural gas
DAB	Dimethyl ammonia borane
DE	Dimethyl ether
DFAFC	Direct-formic acid fuel cell
DLFC	Direct liquid fuel cell
DME	Dimethyl ether
DMFC	Direct methanol fuel cell
DNV	Det Norske Veritas
DNV-GL	Det Norske Veritas-Germanischer Lloyd
DOE	Department of energy
ECF	European climate foundation
ECSA	Electrochemical surface area
EE	Energy efficiency

EG	Ethylene glycol
EGR	Exhaust gas recirculation
EMS	Energy management system
EMSA	European maritime safety agency
ESS	Energy storage system
EU	European Union
EV	Electric vehicle
FA	Formic acid
FC	Fuel cell
FCEV	Fuel cell electric vehicle
FCHEV	Fuel cell hybrid electric vehicle
FCV	Fuel cell vehicle
FESS	Flywheel ESS
FF	Fossil fuel
GDL	Gas diffusion layer
GDP	Gross domestic product
GHG	Greenhouse gases
GVWR	Gross vehicle weight rating
HB	Hydrazine borane
HC	Hydrocarbon
HCNG	Blend of natural gas and hydrogen in variable proportion by volume
HDV	Heavy-duty vehicle
HE	Hydrogen energy
HER	Hydrogen evolution reaction
HEV	Hybrid electric vehicle
HFC	Hydrogen and fuel cell
HHO	Oxy-hydrogen
HPA	Heteropolyacid
HRR	Heat release rate
HRS	Hydrogen refueling station
ICE	Internal combustion engines
ICEV	ICE vehicle
IEA	International energy agency
IMO	International maritime organization
IPPC	Intergovernmental panel on climate change
LCV	lower calorific value
LDV	Light-duty vehicle
LGDL	Liquid gas diffusion layer

LHV	Lower heating value
LNG	Liquified natural gas
LOHC	Liquid organic hydrogen carrier
LPDI	Low-pressure direct cylinder injection
LPG	Liquefied petroleum gas
LT	Low-temperature
MCFC	Molten carbonate fuel cell
MEA	Membrane-electrode assembly
MENA	Middle East and North Africa
MFC	Microfluidic cell
MH	Metal hydride
MOF	Metal-organic framework
MPG	Miles per gallon
MPGe	Miles per gallon gasoline-equivalent
MTPA	Million tonnes per annum
NCMS	National center for manufacturing science
NG	Natural gas
NSGA-II	Non-dominated sorting generating algorithm-II
O&M	Operation and maintenance
OER	Oxygen evolution reaction
ORC	Organic Rankine cycle
ORR	Oxygen reduction reaction
PAAM	Polyacrylamide
PAC	Polymer acid complex
PAES	Poly (arylene ether sulfone)
PAFC	Phosphoric acid fuel cell
PANi	Polyaniline
PAP	Polyarylene piperidines
PBI	Polybenzimidazole
PEC	Photoelectrochemical
PEI	Polyethyleneimine
PEK	Poly(ether ketones)
PEM	Proton exchange membrane
PEME	Proton exchange membrane electrolyzer
PEMFC	Proton exchange membrane fuel cell
PEO	Polyethylene oxide
PESA-II	Pareto envelope-based selection algorithm-II
PGS	Power generation system
PM	Particulate matter

PPy	Polypyrrole
PSS	Polystyrene sulfonate
PtA	Power-to-ammonia
PV/T	Photovoltaic-thermal
PVA	Polyvinyl alcohol
RES	Renewable energy sources
RH	Relative humidity
RORC	Regenerative organic Rankine cycle
RPM	Revolutions per minute
SAXS	Small angle X-ray scattering
SI	Spark ignition
SiC	Silicon carbide
SMR	Steam methane reforming
SOFC	Solid oxide fuel cell
SPE	Solid polymer electrolyte
SPEEK	Sulfonated polyether ether ketones
SPI	Sulfonated polyimides
SPPBP	Sulfonated poly (4-phenoxybenzoyl-1,4-phenylene)
SPPS	Sulfonated poly (phenylene sulfide)
SPSU	Sulfonated polysulfones
SWHC	Soil water holding capacity
TEG	Thermoelectric generator
TLP	Tension leg platform
TMI	Timed manifold injection
TOF	Turn over frequency
TON	Turnover number
TOPSIS	Technique for order preference by similarity to an ideal solution
UAV	Unmanned aerial vehicle
UHC	Unburned hydrocarbon
WGS	Water–gas shift
WOT	Wide-open throttle

1

Introduction to Hydrogen Fueled Power Generation Systems

Onkar Singh[1], Anoop Kumar Shukla[2], Ali J. Chamkha[3], and Meeta Sharma[4]

[1]VMSB Uttarakhand Technical University, India
[2]Department of Mechanical Engineering, Amity University, India
[3]Dean of Engineering, Kuwait College of Science and Technology, Qatar
[4]Department of Mechanical Engineering, Amity University, India.

Abstract

Hydrogen is a clean fuel with minimal impact on the environment and ecology. Its availability from the hydrogen carrying items and its conversion into water upon oxidation in exothermic reaction makes it a potential substitute to the fossil fuels. The concerns for protecting environment and sustainability of flora and fauna make it inevitable to look for clean energy sources. Hydrogen has emerged as a solution to these problems. This chapter details the prospects of hydrogen as a fuel for power generation. A brief discussion about the methods for getting hydrogen and methodologies for using it for power generation is presented herein. It also describes the challenges posed in vast utilization of hydrogen in power generation and mobility solutions.

Keywords: Power Generation, Hydrogen Production, Pyrolysis, Reforming

1.1 Introduction

Present civilization is energy intensive, and all activities are revolving around the energy availability. The pattern of production and consumption depends on the energy that is obtained from burning of fossil fuels. The concerns about regulating carbon footprints are visible across the world and necessitate shifting to green energy options.

It dates back to 1990s when the researchers started looking at hydrogen production and using it for power generation as a potential contributor to decarbonize power generation routes. Hydrogen is zero carbon substitute for existing power generation options. It is an essential ingredient in fertilizer production, petrochemical processing, steel manufacturing, etc., and is likely to be extensively used for power transmission, transportation, etc., in times to come. The high energy density of hydrogen being 120–142 MJ/kg, ease of availability, and clean nature make it suitable to substitute fossil fuels in the future. Burning of hydrogen being devoid of CO_x and soot emissions along with its generation through electrolysis using electrical energy and water make it environment friendly [1].

Various technologies exist today that can produce, store, and transport hydrogen while minimizing costs and the environmental impact during their entire life cycles. Hydrogen can be produced by a variety of processes that emit carbon in different ways, depending on the methods and materials used. Around the world, 120 million tons of hydrogen are produced annually, with just a third being pure hydrogen. Nearly 95% of the hydrogen produced is created using natural gas and coal. About 5% of the total is made up of electrolysis products. The use of hydrogen for getting power for transportation and electricity generation has been successfully attempted through different technological routes. But hydrogen as an energy carrier has certain inherent limitations with regard to storage and transportation.

1.2 Hydrogen Production

There are numerous routes for hydrogen production that rely on extraction of H_2 from hydrogen-containing molecules in water or fossil fuels. Getting hydrogen from ways involving smaller carbon footprint is critical. As per the IEA report [2], out of the total demand for hydrogen, around 95% of 90 million tons/year is obtained from coal gasification or steam methane reforming, which are fossil fuel driven emission intensive routes with the cost of hydrogen production being 0.5−1.7 US$ per kg of H_2. This cost of hydrogen production rises to 1.0−2.0 US$ per kg of H_2 when combined with carbon capture and storage. Different ways for getting hydrogen include its production from renewable and non-renewable sources including fossil fuel or biomass based, microbial hydrogen production, electrolysis, and thermolysis of water and thermochemical cycles [3]. But the challenge of having sustainable, cost-effective, and green route for hydrogen production is to be taken care for its widespread use. Among different routes (Table 1.1),

Table 1.1 Methods of hydrogen generation [13].

Method	Process	Raw material	Energy	Emissions
Thermal	Reformation	Natural gas	Steam and high temperature	CO_2
	Thermochemical hydrolysis	Water	Heat from nuclear energy	No emissions
	Gasification	Coal and biomass	Steam, oxygen, heat, and pressure	Few
	Pyrolysis	Biomass	Steam at medium temperature	Few
Electrical	Electrolysis	Water	Electricity	Varies
	Photo electrochemical	Water	Sunlight	No emissions
Biological	Photobiological	Water and algae	Sunlight	No emissions
	Anaerobic digestion	Biomass	Heat	Few
	Fermentation	Biomass	Heat	Few

the bio-hydrogen production is said to be reliable, stable, efficient, and having a good generation rate. The key hydrogen production methods are as follows [4].

1.2.1 Pyrolysis

Pyrolysis is a chemical process that involves the decomposition of organic materials by heating them in the absence of oxygen. During pyrolysis, the material is subjected to high temperatures (typically between 400 and 800 °C) in a closed container, where it breaks down into smaller molecules such as gases, liquids, and solids. The procedure of producing hydrogen through pyrolysis (Figure 1.1) uses the heat breakdown of organic waste or biomass to produce hydrogen. In the procedure, organic material is heated in a reactor at high temperatures (usually 700–1000 °C) in the presence of steam. Steam combines with the material to produce hydrogen gas (H_2) and other by-products such as carbon dioxide (CO_2), carbon monoxide (CO), and methane (CH_4). Since pyrolyzed hydrogen burns cleanly and only releases water vapor when it is burned, it can be utilized as a fuel for both transportation and power generation. Moreover, hydrogen can be used as a raw material in the synthesis of compounds like methanol and ammonia. As it may use a variety of organic resources, including biomass, municipal trash, and agricultural waste, pyrolysis can be a desirable method for producing hydrogen. To provide a reliable source of hydrogen, the process can also be combined with other renewable energy sources like solar and wind energy.

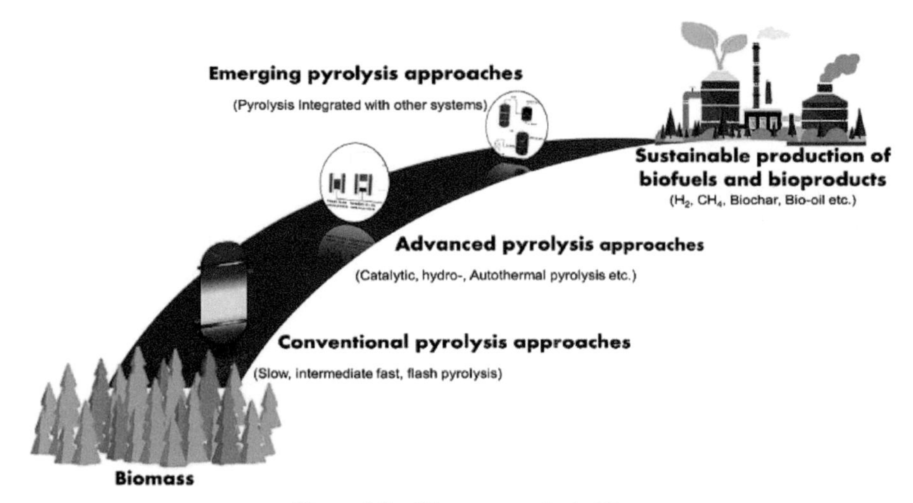

Figure 1.1 Biomass pyrolysis [5].

The procedure does present certain difficulties, though, such as the requirement for high temperatures and the potential creation of dangerous by-products. The quality and composition of the organic material utilized as a feedstock might also have an impact on the process's efficiency. Overall, the pyrolysis method for manufacturing hydrogen is a potential way to provide clean and renewable fuel, but more research and development are required to improve the method and guarantee its environmental sustainability.

1.2.2 Gasification and reforming

Hydrogen production from water gas comprising hydrogen, carbon monoxide, and carbon dioxide using fixed-bed gasifier is quite common and around 18% of hydrogen requirement is met from coal-derived hydrogen, which has been prevalent since 1833. Natural gas and oil could also be used for getting hydrogen through syngas method. As sources for hydrogen production, the natural gas and liquid hydrocarbons amount to 48% and 30% of total hydrogen sources. Water gas shift reaction is required for getting hydrogen.

Production of hydrogen from biomass through gasification route is an attractive way (Figure 1.2). The forest residue, agricultural waste, municipal waste, etc., can be used for economic production of hydrogen through gasification technology using suitable gasification agents. Amongst different thermochemical processes used for hydrogen production, the steam gasification produces the maximum of hydrogen. However, the type of biomass,

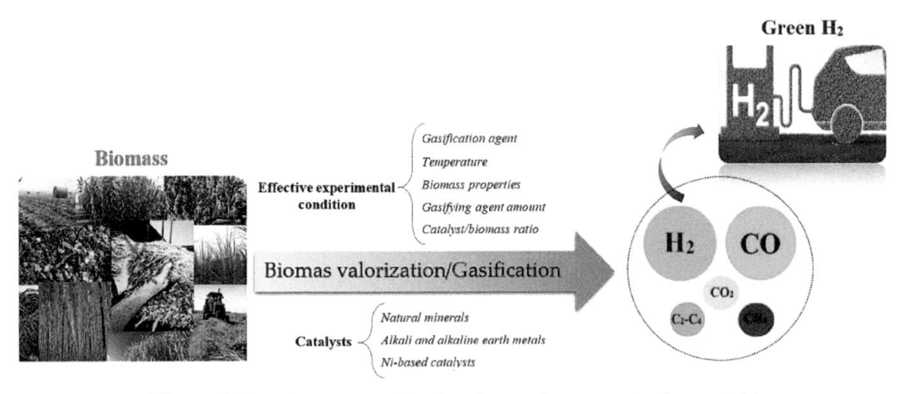

Figure 1.2 Biomass gasification for getting green hydrogen [6].

feed quality, solvent, temperature/pressure during reaction, catalyst, ratio of sorbent to biomass, residence time, etc., influence the yield of hydrogen in gasification.

For example, the hydrogen production from methane is governed by the following set of reactions:

$$CH_4 + 2O_2 \rightarrow 2\,H_2O \;+\; CO_2.$$

Endothermic reaction yielding hydrogen:

$$CH_4 + H_2O -> 3H_2 + CO$$

$$CO + H_2O \rightarrow H_2 + CO_2.$$

Combining the above equations, the vapor reforming of methane occurs at moderate pressure of 2–3 MPa and high temperature of around 1200 K.

$$CH_4 + 2H_2O \rightarrow 4H_2 + CO_2\,.$$

Methane can also be oxidized to hydrogen through faster reaction rate; however, in the case of partial oxidation, the cost is quite high.

$$2CH_4 + O_2 \rightarrow 4H_2 + 2CO.$$

1.2.3 Photosynthesis

The process of producing hydrogen through photosynthesis involves splitting water molecules (H_2O) into their component elements of hydrogen (H_2) and oxygen (O_2) using sunlight (O_2). Photosynthetic pathway yields hydrogen using solar energy. This is also termed as biological hydrogen production and requires smaller temperature in range of 10–40 °C in the presence of hydrogen donor like salt water. Temperature requirement is small as compared to chemical/physical routes of hydrogen production requiring temperatures of the order of 1000 K. Such production is clean. The employment of specific photocatalysts, such as titanium dioxide, that can absorb sunlight and use that energy to start the photolysis of water molecules is the most promising method for producing hydrogen through photosynthesis. These photocatalysts are often incorporated into a system that also contains a hydrogen collection system and a water source (such as a water tank or membrane).

One benefit of producing hydrogen through photosynthesis is that it makes use of abundant and renewable resources (sunlight and water) to create a sustainable and clean fuel. Also, the method is environmentally benign because it does not emit any toxic gases. This method still faces significant difficulties, such as the relatively poor photocatalyst efficiency, the requirement for specific materials, and the high implementation costs. Despite these difficulties, researchers are still looking into and developing this promising hydrogen generating technique.

1.2.4 Water electrolysis

Water electrolysis is a process in which hydrogen gas (H_2) is produced by splitting water (H_2O) into its constituent elements of hydrogen and oxygen using an electric current. This process involves passing an electric current through water, which causes the water molecules to split into hydrogen and oxygen ions.

In order to electrolyze water, a device called an electrolyzer is commonly used. It has two electrodes (an anode and a cathode) that are separated by an electrolyte solution. Hydrogen ions (H^+), which are drawn to the cathode when an electric current is run through the electrolyte solution, interact with the cathode's electrons to produce hydrogen gas. At the same time, oxygen ions (O^{2-}) are drawn to the anode and join forces with anode electrons to create oxygen gas. It is found that for generating 1 kg of hydrogen (Figure 1.3), there is a requirement of electrolysis of 9 L of water and around 50 kWh in an electrolyzer working with 70% electrical efficiency [1].

The common electrolyzers used for hydrogen production include solid oxide electrolyzer cell (SOEC), polymer electrolyte membrane electrolyzer (PEM), and alkaline electrolyzer [1]. Depending on variables like efficiency, cost, and scalability, each style has advantages and cons of its own. Nejadian

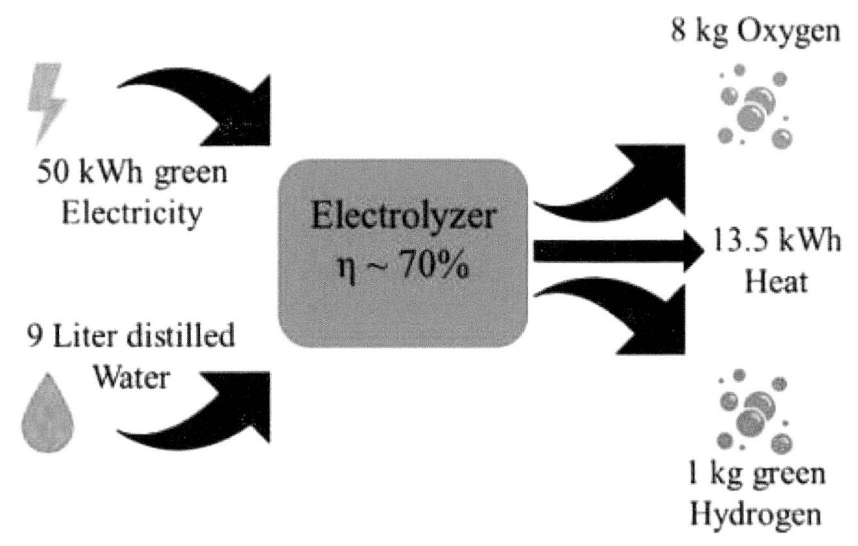

Figure 1.3 Electrolysis of water for producing hydrogen [1].

et al. [8] showed that amongst these, SOEC works at high temperatures and offers the maximum hydrogen production rate at the cost rate of 0.9372 $/s, while the highest cost rate of 3.54 $/kg is found in the PEM system, and the cheapest is from the alkaline system at 2.94 $/kg.

One benefit of producing hydrogen by water electrolysis is that the necessary electric current can be produced using sustainable energy sources like solar and wind power [7]. The method is also environmentally benign because it does not emit any toxic gases. However, there are still some difficulties with this method, including the relatively high cost of installing electrolysis devices, the demand for highly pure water, and the need for substantial amounts of electricity to generate considerable amounts of hydrogen. Notwithstanding these difficulties, water electrolysis is a promising method for producing hydrogen, and more research and development are anticipated to result in even greater gains in effectiveness and cost-savings.

1.3 Hydrogen in Power Generation

Petroleum can be replaced with hydrogen fuel in several industries as fuel for power generation with fuel cell systems. Today, steam methane reforming accounts for more than 95% of all hydrogen generation, and it is strongly encouraged to manufacture green hydrogen by water electrolysis. In addition, industrial waste streams with a high hydrogen content can yield hydrogen with only a preliminary purification step. Using a fuel cell system, this hydrogen is used to balance the grid when necessary. It is also blended with the natural gas grid and used as feedstock in industrial processes in steelmaking, chemical, and refinery plants (power-to-gas). Hydrogen is also used as fuel in the transportation sector. Due to its adaptability, hydrogen is a very compelling option, and because of this, important nations and regions including India, Canada, Australia, USA, Japan, and Europe have established several development roadmaps and strategies. Nowadays, more research on how hydrogen is used is being conducted, particularly in the power sector, and how it may support a variety of power cycles, such as those in hydrogen-fired, integrated gasification combined cycle, and co-combustion power plants.

Oxy-hydrogen (HHO) fuel, which is also called Brown's gas, can be used for heating and power generation. HHO holds the potential to run internal combustion engines, boilers, welding, cutting, etc., applications wherever larger amount of heat is required. Figure 1.4 shows the potential of HHO as an energy source.

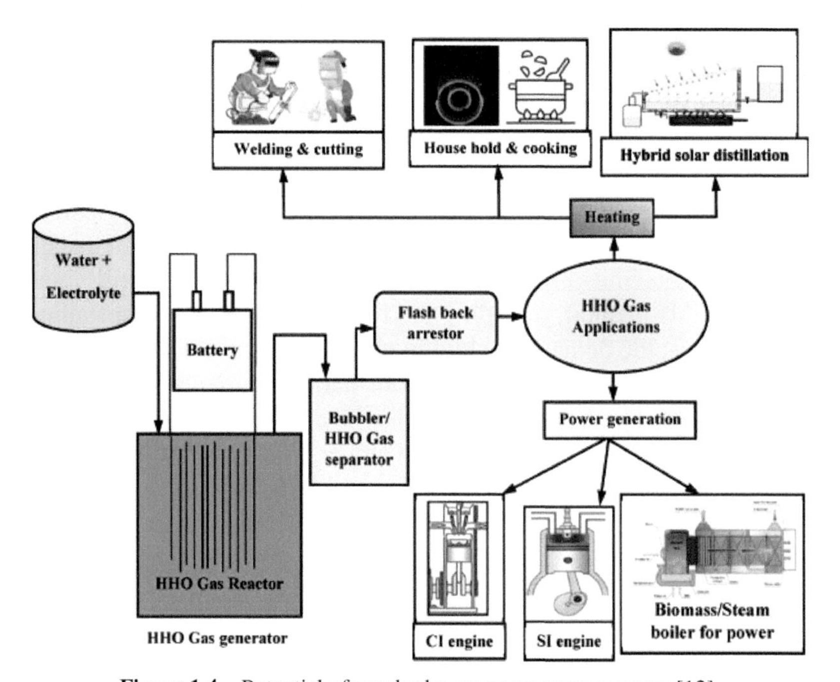

Figure 1.4 Potential of oxy hydrogen as an energy source [12].

1.3.1 Co-combustion (natural gas–hydrogen mix)

Existing natural gas-fired power stations could use hydrogen as a means of decarbonization. Global natural gas-fired generating capacity stood at about 1900 GW as of 2020. Natural gas and hydrogen mixtures are generally compatible with CCGTs in use. However, the maximum amount of hydrogen that can be present in the mixture varies depending on the make and type of the turbine [9]. The challenges brought on by hydrogen's greater flame temperatures, faster laminar flame, and shorter auto-ignition delay compared to natural gas will be mitigated by new turbine designs and materials. Given that these power plants reduce GHG emissions, enhance fuel variety, and encourage technical advancement, nations may decide to commission them.

Coal-fired power stations might decarbonize with hydrogen as well. In Japan, experiments are being conducted to see if co-firing coal and the hydrogen-carrying ammonia it contains is technologically and economically feasible. Global coal-fired capacity was 2150 GW as of 2020 [9]. If green ammonia is employed, co-firing coal with ammonia up to 20% by energy content will cut these facilities' annual CO_2 emissions by almost 1.7 billion t. [9]. Co-firing would enable coal-fired power facilities to continue operating, despite the lack of viable decarbonization options. However, coal-fired power facilities all around the world are being pushed to shut down earlier than their design life permits, either by market forces or government legislation. As a result, co-firing ammonia and coal could be a transition fuel.

1.3.2 Blended hydrogen fuel IC engine

For quite some time, the simultaneous pressures of energy conservation and environmental preservation are making it essential to identify new clean energy sources and all attempts are made to replace conventional internal combustion engines with some suitable alternatives. As a consequence, the hydrogen is seen as a potential source of energy that offers clean and efficient combustion. Also, the regenerative capabilities of hydrogen demonstrate possibilities of its becoming a replacement to the conventional fossil fuels as an excellent fuel for internal combustion engines. Hydrogen fuel has a very low energy of ignition compared to other conventional fossil fuels. During the combustion of hydrogen, the flame diffuses very fast and also has small quenching distance. This also helps in better homogeneity of the combustibles, short duration of combustion, increased engine power, better economy and stability, and less harmful emissions. In certain cases, the internal combustion engines employ hydrogen as a fuel extender by mixing it to gasoline/diesel for deriving the benefits of hydrogen combustion [14].

Literature shows that there are several studies that report about the research works on reciprocating engines and rotary engines that run on gasoline blended with hydrogen. Also hydrogen blending to natural gas and alcohol is separately tried in different types of engines. Ji et al. [15] performed experimental studies on the Z160F rotary engine's combustion and emissions employing hydrogen blending in gasoline. The study claimed that after H_2 enrichment, the pressure of combustion, cylinder temperature, brake mean effective pressure, and thermal efficiency all rose. They also discovered that CO emissions fell after H_2 blending and that HC emissions declined by 44.8% when 5.2% hydrogen (volume percentage) was added. The effects of the quantity of H_2 addition (ranges from 6% to 18% by volume) on the combustion process and NO production in a hydrogen-blended diesel engine were examined by Wang et al. [17] using a numerical simulation model. According to simulation results, the ignition delay was found to be longer, premixed combustion was found to be improved, and the diffusion of diesel combustion was found to be encouraged after H_2 addition. Additionally, they reported that the injection of H_2 might reduce soot emissions and limit the rise in NO emissions (as shown in Figure 1.5).

In recent years, some researchers have combined hydrogen with other low heat value gases, such as dimethyl ether (DME), methane (CH_4), or biogas (primarily CNG, N_2, and CO_2) and used the resulting mixtures as fuels in engines in addition to using hydrogen as an additional fuel in gasoline, diesel, natural gas, and alcohol engines [14]. Kekec and Karyeyen [17] looked at how CH_4–H_2 blending fuels (i.e., 60% CH_4 – 40% H_2, 50% CH_4 – 50% H_2, and 60% CH_4 – 60% H_2) burned and released pollutants under standard and color-less distributed combustion (CDC) circumstances in a cyclonic combustor. At a specific oxygen content, they found that color-less distributed combustion resulted in a sizable reduction in NOx and a beneficial reduction in CO.

Worldwide research works show significant advancements and achievements in hydrogen blending to different fuels such as gasoline, diesel, alcohol, and natural gas for various types of engines. The effects of H_2 blending on the performance of combustion, emission, output from these engines, implementation of lean combustion, innovations and control strategy optimization, studying the influence of various process variables on the performances, emission standards of the engines, etc., are the main components of this advancement.

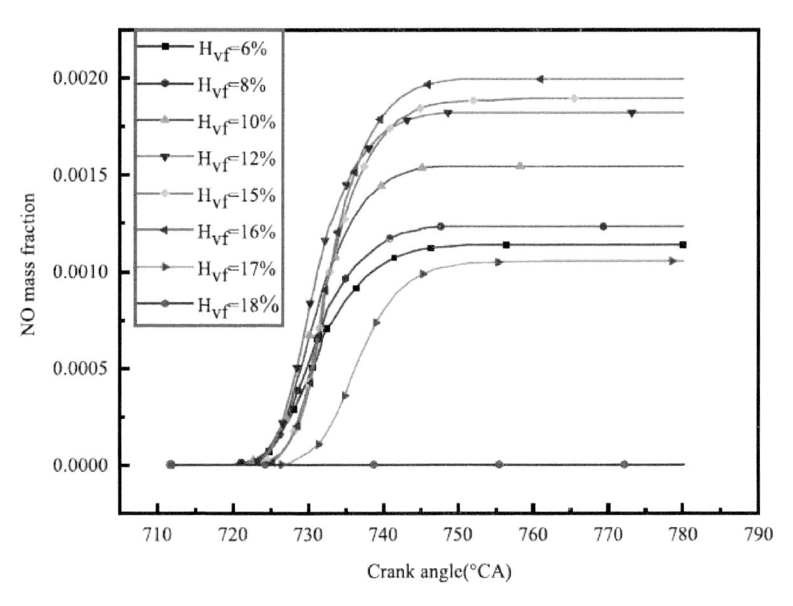

Figure 1.5 Variations of NO emissions at different volume fractions of H_2 in a hydrogen-blended diesel engine [16].

1.4 Global Trends for Green H_2 Production

With the help of goals, policies, and strategies defined by many nations, the market for decarbonized hydrogen is growing. As of August 2022, 38 nations and the European Union (EU) had declared decarbonized hydrogen policies with explicit goals for increasing electrolyzer capacity and producing hydrogen, according to CEEW's study [18]. Among them, 29 are advanced economies. The EU intends to install 40 GW of electrolyzer capacity by 2030 in order to produce 10 MTPA of green hydrogen. For this, investments in electrolyzers and renewable energy will total USD 46 billion and USD 520 billion, respectively. China has set a goal of adding 80 GW of electrolyzer capacity by 2060, which is the highest in Asia. With a goal of 0.42 MTPA by 2030, Japan has begun harnessing solar energy in Fukushima to produce green hydrogen. By 2030, India's National Green Hydrogen Mission, which was first announced in 2021, hopes to add 5 MPTA of green hydrogen. Major oil producers in the Middle East and North Africa (MENA) area are diversifying their energy portfolios. The largest green ammonia plant in the world is being built in Saudi Arabia, using 4 GW of renewable energy capacity for electrolysis to create 1.2 MTPA of green ammonia. Africa's contribution to the world's total renewable energy capacity in 2020 was only 54 GW, or less than 2%. Only seven of the 604 low-carbon hydrogen projects listed in the International Energy Agency's (IEA) database are being developed on the African continent: in South Africa, Mauritania, Egypt, and Morocco. The only African nations with specified green hydrogen targets are Namibia and South Africa. With lofty goals of 5- and 25-GW electrolyzer capacity increase by 2025 and 2030, respectively, Chile has assumed leadership in Latin America. The 3-GW target for Colombia by 2030 is similar to that of advanced nations like Austria, the Netherlands, and Portugal.

1.5 Challenges with Hydrogen as Fuel

Along with several advantages of hydrogen fuel is the process of decarbonizing the environment. Given the differences between hydrogen and many conventional hydrocarbon fuels, there are hurdles that must be acknowledged even though using hydrogen can result in lower CO_2 emissions. There are several challenges that need to be addressed for utilizing hydrogen as a fuel in power generation. Hydrogen is two times more energetically dense than methane on a mass basis (Table 1.2). However, the comparison of energy per unit volume between hydrogen and methane shows that the latter has three times more than the former. As a result, hydrogen requires a flow of three times the volume to produce the same amount of heat (energy) as methane. Therefore, a fuel accessory system setup for the requisite flow rates is needed to operate a gas turbine on 100% hydrogen [10].

With hydrogen, there are extra operational difficulties that concern general security. First off, a hydrogen flame is dim and challenging to view

Table 1.2 Lower heating value and flame speed of fuels [10, 11].

Name of fuel	Chemical formula	Lower heating value (LHV)	Laminar flame speed at stoichiometric conditions
Hydrogen	H_2	120.0 MJ/kg 10.8 MJ/m^3	170 cm/s
Methane	CH_4	50.0 MJ/kg 35.8 MJ/m^3	38.3 cm/s
Ethane	C_2H_6	47.8 MJ/kg 27.3 MJ/m^3	40.6 cm/s
Propane	C_3H_8	46.4 MJ/kg 23.1 MJ/m^3	42.3 cm/s
Ethanol	C2H6O	26.7 MJ/kg 21.1 MJ/m^3	45.5 cm/s

with the naked eye. This necessitates the use of flame detection devices designed specifically for hydrogen flames. Second, despite being thought of as airtight or impenetrable to other gases, hydrogen can seep through seals. As a result, welded connections or other suitable parts may need to be utilized in place of conventional sealing systems used with natural gas. Third, the lower flammability limit for hydrogen is 4%, whereas it is 5% for methane (in air). As a result, hydrogen leaks could lead to higher safety concerns, necessitating modifications to industrial practices, safety/exclusion zones, etc. There might also be other plant-level safety issues that need to be looked at.

The storage of hydrogen is also one of the major deterrents for its use in certain applications of power generation. The underground and overground storage options have to be integrated for realizing the hydrogen energy chain. Its large quantity storage in salt caverns, unserviceable hydrocarbon reservoirs, aquifers, hard rock caverns, etc., are being attempted; however, it requires huge structures depending upon the storage cycles, capacities, and hydrogen purity needed at the time of its retrieval. The overground storage of hydrogen is good for applications that are of stationary or mobile nature needing small capacity and fast storage cycle. The storage of hydrogen is done in different forms like compressed, cryogenic, or its combination in the pressure vessels. Also, there is a possibility of storing hydrogen by adsorbing or absorbing it in different materials; however, its selection depends upon the specific application for which hydrogen is the energy source.

The presence of hydrogen in different fuels, whether in small or large quantity, makes them suitable for being used to run gas turbines, but the hydrogen alone has extraordinary capabilities and environment friendliness for powering these gas turbines. Undoubtedly, there are various difficulties in the use of hydrogen as fuel in gas turbines and other engines, which need to be sorted out in due course of time. Many of the technical problems regarding the suitability of this fuel for power generation applications have been resolved as a result of industrial experience with hydrogen-based fuels. Consequently, it is important to think of retrofitting of current gas turbine power facilities as an integral part of any future power to hydrogen ecosystem.

References

[1] Marcel Otto, Katerina L. Chagoya, Richard G. Blair, Sandra M. Hick, Jayanta S. Kapat, Optimal hydrogen carrier: Holistic evaluation of hydrogen storage and transportation concepts for power generation, aviation, and transportation, Journal of Energy Storage, Volume 55, Part

D, 2022, 105714, ISSN 2352-152X, https://doi.org/10.1016/j.est.2022.105714.

[2] IEA (2021), Global Hydrogen Review 2021, IEA, Paris https://www.iea.org/reports/global-hydrogen-review-2021.

[3] Mengdi Ji, Jianlong Wang, Review and comparison of various hydrogen production methods based on costs and life cycle impact assessment indicators, International Journal of Hydrogen Energy, Volume 46, Issue 78, 2021, Pages 38612-38635, ISSN 0360–3199, https://doi.org/10.1016/j.ijhydene.2021.09.142.

[4] R. Yukesh Kannah, S. Kavitha, Preethi, O. Parthiba Karthikeyan, Gopalakrishnan Kumar, N. Vo. Dai-Viet, J. Rajesh Banu, Techno-economic assessment of various hydrogen production methods – A review, Bioresource Technology, Volume 319, 2021, 124175, ISSN 0960-8524, https://doi.org/10.1016/j.biortech.2020.124175.

[5] Arun Krishna Vuppaladadiyam, Sai Sree Varsha Vuppaladadiyam, Abhishek Awasthi, Abhisek Sahoo, Shazia Rehman, Kamal Kishore Pant, S. Murugavelh, Qing Huang, Edward Anthony, Paul Fennel, Sankar Bhattacharya, Shao-Yuan Leu, Biomass pyrolysis: A review on recent advancements and green hydrogen production, Bioresource Technology, Volume 364, 2022, 128087, ISSN 0960-8524, https://doi.org/10.1016/j.biortech.2022.128087.

[6] Valizadeh, S., Hakimian, H., Farooq, A., Jeon, B., Chen, W., Hoon Lee, S., Jung, S., Won Seo, M., & Park, Y. (2022). Valorization of biomass through gasification for green hydrogen generation: A comprehensive review. *Bioresource Technology*, *365*, 128143. https://doi.org/10.1016/j.biortech.2022.128143

[7] Xianxian Xu, Quan Zhou, Dehai Yu, The future of hydrogen energy: Bio-hydrogen production technology, International Journal of Hydrogen Energy, Volume 47, Issue 79, 2022, Pages 33677-33698, ISSN 0360-3199, https://doi.org/10.1016/j.ijhydene.2022.07.261.

[8] Mehrnaz Mohebali Nejadian, Pouria Ahmadi, Ehsan Houshfar, Comparative optimization study of three novel integrated hydrogen production systems with SOEC, PEM, and alkaline electrolyzer, Fuel, 2022, 126835, ISSN 0016–2361, https://doi.org/10.1016/j.fuel.2022.126835.

[9] "Hydrogen's Role in Power Generation | Argus Media" [Online]. Available: https://www.argusmedia.com/en/blog/2021/april/20/hydrogens-role-in-power-generation. [Accessed: 24-Jan-2023].

[10] Goldmeer, J. (2019). POWER TO GAS: HYDROGEN FOR POWER GENERATION Fuel Flexible Gas Turbines as Enablers for a Low or

Reduced Carbon Energy Ecosystem. GEA33861, 1–19. Retrieved from https://www.ge.com/content/dam/gepower/global/en_US/documents/fuel-flexibility/GEA33861PowertoGas-HydrogenforPowerGeneration.pdf

[11] Gülder, Ö. L. (1982). Laminar burning velocities of methanol, ethanol and isooctane-air mixtures. Symposium (International) on Combustion, 19)1), 275–281. https://doi.org/10.1016/S0082-0784(82)80198-7

[12] Paparao, J., & Murugan, S. (2021). Oxy-hydrogen gas as an alternative fuel for heat and power generation applications - A review. International Journal of Hydrogen Energy, 46(79), 37705–37735.

[13] Bairrão, D., Soares, J., Almeida, J., Franco, J. F., & Vale, Z. (2023). Green Hydrogen and Energy Transition: Current State and Prospects in Portugal. Energies, 16(1), 551.

[14] Wang, L., Hong, C., Li, X., Yang, Z., Guo, S., & Li, Q. (2022). Review on blended hydrogen-fuel internal combustion engines: A case study for China. Energy Reports, 8, 6480–6498.

[15] Ji, C., Su, T., Wang, S., Zhang, B., Yu, M., & Cong, X. (2016). Effect of hydrogen addition on combustion and emissions performance of a gasoline rotary engine at part load and stoichiometric conditions. Energy Conversion and Management, 121, 272–280.

[16] Wang, Lijun, Liu, Dong, Yang, Zhenzhong, Li, Hailin, Wei, Leyu, Li, Quancai, 2018. Effect of H2 addition on combustion and exhaust emissions in a heavy-duty diesel engine with EGR. Int. J. Hydrog. Energy 43, 22658–22668.

[17] Kekec, K. B., & Karyeyen, S. (2022). H2–CH4 blending fuels combustion using a cyclonic burner on colorless distributed combustion. International Journal of Hydrogen Energy, 47(24), 12393–12409.

[18] Ghosh, A., Gupta, T., Raha, S., Mallya, H., Yadav, D., and Harihar, N., 2022, "Hydrogen Decarbonisation | Rules for an Energy-Secure Green Economy," Counc. ENERGY, Environ. WATER [Online]. Available: https://www.ceew.in/publications/energy-secure-green-hydrogen-economy-and-decarbonisation. [Accessed: 13-Feb-2023].

2

Polymer Electrolyte Membrane Fuel Cell (PEMFC) Membranes

Arzu Göbek[1] and Ayşe Bayrakçeken Yurtcan[2,3]

[1]Faculty of Science, Department of Chemistry, Atatürk University, Turkey
[2]Faculty of Engineering, Department of Chemical Engineering, Atatürk University, Turkey
[3]Graduate School of Science, Department of Nanoscience and Nanoengineering, Atatürk University, Turkey
E-mail: arzugbk89@gmail.com; ayse.bayrakceken@gmail.com

Abstract

Polymer electrolyte membrane fuel cells (PEMFCs) use an electrolyte that is based on a polymer membrane and has special properties that allow it to conduct protons without letting the passage of electrons or gases. This electrolyte is located between two electrodes. It must have high proton conductivity and be stable at high temperatures and in chemicals. Most of the polymer-based membranes used in PEMFCs operate at low temperatures. The biggest obstacle to getting the best efficiency from a PEMFC operating with reformate gases is CO poisoning. The presence of a certain amount of CO in the environment leads to a significant decrease in the efficiency of the anode catalyst. Operating PEMFC at high temperatures has recently been seen as a way to both overcome this issue and increase the kinetics of the reactions. Therefore, a PEMFC membrane should have good thermal stability and mechanical strength at high temperatures, ideal thickness, high ionic conductivity, low gas permeability, and low cost. PEMFC membranes are classified as either proton exchange membranes (PEMs), in which negative ions are retained at the backbone ends of a polymer mold and this charge is balanced with protons, or as polymer acid complexes (PACs), where a basic polymer is loaded with acidic components. In this chapter, the physical, thermal, and mechanical properties, ion-conducting principles, and characterization methods of cationic PEMFC membranes operating at low and high temperatures will be examined. The addition of different materials in order to improve the properties of the membranes will also be covered in this chapter.

Keywords: Membrane, polymer, proton conductivity, PEMFC

2.1 Introduction

Due to its many benefits, such as being clean, quiet, modular, efficient, portable, sustainable, reliable, and having low emissions (Tawil et al., 2007), fuel cell technology is being used widely as a way to the transition to clean energy. A fuel cell is an electrochemical energy converter that can convert fuel from chemical energy to electrical energy in just one step. Fuel cells are very advantageous compared to traditional energy conversion systems that generate electricity. These systems can be used as ideal power generators as well as compact, flexible, quiet, efficient, and environmental-friendly systems that can be used for different purposes (Ergün, 2009).

The substance between the electrodes of fuel cells, known as the electrolyte, is typically what distinguishes them. In the literature, it was stated that the best temperature to run the fuel cell and the best fuel to generate electricity depend on the properties of the electrolyte (Ergün, 2009). Even though different types of fuel cells operate similarly, there are differences in the reactions that happen at the anode and cathode electrodes, the types of electrolytes (membranes) used, the working conditions, and the areas where they can be used.

Ion exchange membranes are divided into two groups: anion exchange and cation exchange membranes. These membranes are classified according to the positive or negative ion groups connected to the membrane matrix. Membranes in which negative ions such as SO_3^-, COO^-, PO_3^{2-}, PO_3H^-, $C_6H_4O^-$, etc., are fixed to the membrane backbone are cation exchange membranes that permit the transition of cations and prevent the transition of anions. Membranes in which positive ions like NH_3, NRH_2^+, NR_2H^+, NR_3^+, PR_3^+, and SR_2^+ are fixed to the membrane backbone are called anion exchange membranes. These types of membranes permit the passage of anions but prevent the passage of cations (Xu, 2005). The way that charge groups are attached to the membrane matrix is another way to divide ion exchange membranes. These membranes are called homogeneous and heterogeneous ion exchange membranes. The charge groups are either attached to the membrane matrix by chemical bonds or physically blended with the membrane matrix. The homogeneous ion exchange membrane has relatively better electrochemical properties, but its mechanical strength is not good. The heterogeneous ion exchange membrane, on the other hand, is mechanically more durable and has better dimensional stability but has weak electrochemical performance (Peighambardoust et al., 2010).

PEMFC is a kind of fuel cell with an electrolyte made of a solid polymer with proton exchange ability (Lee et al., 2014). These membranes, which are types of ion exchange membranes and the basic components of the fuel cell, are the components that directly affect the properties and operating life of the fuel cell (Chen et al., 2020). PEMFCs have a lot of benefits compared to other types of fuel cells. They can use hydrogen, which is a clean fuel, they do not release pollution, they have a high energy density (especially compared to battery systems), and they have a high conversion efficiency of up to 60%. It is an environmentally friendly system (Çelik et al., 2012).

2.1.1 PEMFC membranes operating at low or high temperature

Wong et al. (2019) say that the electrolyte material for PEMs is chosen based on whether the fuel cells work at low or high temperature. Low-temperature (LT) PEMs operate at temperatures between room temperature and 100 °C, while high-temperature PEMs operate between 120–180 °C (Kurz et al., 2018). Some difficulties have been encountered in developing low-temperature PEM fuel cells used in transportation and mobile device technologies (Wong et al., 2019). The perfluorinated sulfonic acid membranes are used for the PEM fuel cell, which operates at low temperatures in hydrated conditions. Therefore, water management in the LT-PEM fuel cell is quite complex. It is also very important to balance the water content in LT-PEMFCs. Otherwise, water droplets accumulating under the GDL and flow channels may partially block the gas supply of the cell (Zeis, 2015). LT-PEMFCs have a low tolerance for CO and other impurities (Wong et al., 2019). The presence of a certain amount of CO in the environment negatively affects the efficiency of the fuel cell when PEMFCs operate with reformate gases. The CO poisoning of the fuel cell causes a significant reduction in the efficiency of the anode catalyst (Hasiotis et al., 2001). The conductivity of the membrane (Nafion) used in PEMFCs operating at less than or equal to 80 °C decreases as the relative humidity (RH) decreases. Therefore, an external humidification subsystem is needed to provide conductivity at high temperatures. In addition, since the glass transition temperature of the membrane is low (80–120 °C) in PEMFCs, it loses its mechanical and dimensional stability with increasing temperature. Because of these problems, efforts have been made toward increasing the operating temperature of the PEMFCs and because the desired operating temperature for PEMFCs is considered to be 120–150 °C (Shao et al., 2007).

Because of the problems with low-temperature fuel cells that we have already talked about, there has been more interest lately in making high-temperature fuel cells. In contrast to Nafion-based LT-PEMFCs, it has been suggested that HT-PEMFCs could speed up the reaction and protect the catalyst (Jung et al., 2012). Singh et al. (2018) say that the high-temperature fuel cell that works between 150 and 300 °C has given an important direction to fuel cell research because it is easier to handle water and heat, the fuel can be used in different ways, and non-precious metal catalysts can be used. Also, there are many advantages to operating the PEMFC above 100 °C. (i) The oxygen reduction reaction (ORR) occurring at the cathode has slow reaction kinetics. High-temperature operation provides a faster electrochemical reaction on the metal catalyst. (ii) The water released as waste is easily removed by evaporation in the cathodic exhaust stream. (iii) Humidification and water removal processes are easier because PEMFC systems can be operated in dry gas feed streams. (iv) The tolerance of PEMFC systems operating at high temperatures against fuel contaminants is one of their most important advantages (Perry et al., 2014).

Besides the advantages, there are some challenges to operating PEMFCs at high temperatures. Aqueous-based membranes are used in LT-PEMFCs because the proton conductivity is significantly dependent on relative humidity. As the temperature increases, the proton conductivity decreases as a result of dehydration of the membrane (Zhang et al., 2006). In fact, it is quite difficult to find an electrolyte membrane with good stability and conductivity at high temperatures. Because of the high-temperature environment and possible corrosion and mechanical malfunction in bipolar plates, deterioration of the supporting carbon material, PEM, and catalyst or catalyst layer begins to occur. Due to corrosion, the electrochemical surface area (ECSA) decreases rapidly, resulting in poor performance. In addition, corrosion causes stack failure and stack safety as it increases contact resistance. Besides all these, heating strategies are inadequate to continuously operate HT-PEMFC stacks to reach and retain high temperatures. Considering all of these situations, researchers have been trying to reduce these problems and enrich HT-PEMFCs (Rath et al., 2020; Haider et al., 2021).

In PEMFCs, PEM is one of the most important components due to its role in dispersing fuel, transporting air (Rosli et al., 2017), conducting protons, and separating fuel from an oxidizing agent (Wang et al., 2018). In addition, PEMs, which are the main components of PEMFCs, have three important tasks. PEM acts as a barrier to facilitate proton transfer, to prevent electrons from passing through the cell due to the existence of SO_3^- and to prevent the passage of fuels between the anode and the cathode (Harilal et al., 2020; Ogungbemi et al., 2019). PEMs, which are used as electrolytes in PEMFCs, are of great interest today, with unique features such as high chemical stability against oxygen and free radicals, good mechanical elasticity, and high proton conductivity (Pu et al., 2010). Compared to liquid electrolyte systems, solid PEMs have countless advantages, such as ease of use, compact structure, suitability for mass production, great resistance to the permeability of gaseous reactants, and the ability to produce very thin films. In addition, a solid polymer membrane that is tough and flexible is an important structural component that makes it much easier to transport, seal, and install than liquid electrolyte fuel cells (Li et al., 2009).

2.2 Membrane properties

The performance of a PEMFC is directly connected to the properties of the PEMs. For a PEMFC to perform well, a solid polymer electrolyte (SPE) should have properties such as (a) good proton conductivity, (b) low cost, (c) good water uptake, (d) long-term durability, (e) low electronic conductivity and permeability, (f) sufficiently good chemical, electrochemical, thermal, hydrolytic, mechanical, morphological, and dimensional stability, (g) low gas (fuel and oxidant) permeability, and (h) low swelling levels. All these properties of the membrane must be optimized in order to be able to apply them in a fuel cell. For this reason, studies have focused on membranes that can be used in fuel cell applications by modifying the polymeric membrane in order to retain the proton conductivity at high temperatures, besides the performance of the membrane (Brandon et al., 2003; Kraytsberg & Ein-Eli, 2014; Bose et al., 2011).

Main features of a fuel cell membrane:

- ▷ Proton conductivity
- ▷ Water uptake
- ▷ Gas permeability

Physical features of a fuel cell membrane:

▷ Strength
▷ Dimensional stability (Barbir, 2006)

In membranes, it is important to control things like how water is taken in and moved, how gases pass through, and how protons move through. If these parameters are not managed well, different problems may arise. For example, weakness in water management meant that water transportation was also weak. This leads to problems such as drying out or overflowing. All these features of the membrane are very important because the performance and endurance of a membrane are determined by these properties. Also, in order for a material to be called a membrane, it must be durable and strong, have a high interest in protons, and have good resistance to chemical assaults (Ogungbemi et al., 2019).

There are various factors influencing the thermal and mechanical features of PEMs. These are water, resolvent, operating temperature, grade of sulfonation, and filler for composite membranes (Zaidi & Matsuura, 2008). Basically, the electrical conductivity of the membrane in the fuel cell is highly adherent to the water content (Chen et al., 2004). Consequently, the performance of a membrane is linked to the proton conductivity, and the proton conductivity is linked to the hydration level. Another way to improve fuel cell performance is to decrease the membrane thickness. By reducing the membrane thickness, water entrainment and passage can be prevented. Another advantage is that the membrane resistance and cost are low, and hydration is fast (Smith et al., 2005).

According to the studies, the resistance goes up as the thickness of the membrane goes down, but the performance goes up as the thickness goes up. With thinner membranes, the resistance goes down; so the number of protons and the flow go up. However, since reducing the membrane thickness is a negative situation that leads to an increase in fuel crossover, the membrane thickness must be adjusted according to the operating conditions (Ogungbemi et al., 2019; Atifi et al., 2014).

Operating temperature is the most important parameter affecting membrane performance because many activities to achieve a good performance are dependent on temperature. The operating temperature not only affects the conductivity of the membrane but also has a significant effect on properties such as the protonic resistance required for mass transport and gas diffusion. In addition, with the increase in temperature, the rate of the electrochemical reaction and the formation of protons also increase (Ogungbemi et al., 2019).

The overall performance of the PEM in a PEM fuel cell cannot just be judged by how well it moves protons. The cost, durability, and compatibility of PEM with catalyst and electrode materials are also very important factors that should not be overlooked. There is a very close relationship between the durability and chemical stability of a membrane (Zhang et al., 2012). Also, the durability of PEMs has a lot to do with how long fuel cells can work. Kusoglu et al. (2006) say that mechanical damage to the PEM can happen if the membrane thickness is not changed, pinholes form, or the membrane and GDL separate. The physical features and performances of some membrane classes are shown in Table 2.1.

Table 2.1 Physical properties and performance of some polymer types (references in the table are valid).

Membrane class	Example	Physical features	Performance	Ref.
Perfluoro sulfonated membranes	Nafion	– Durable and stable in oxidative and reducing ambients	– Long endurance time – High conductivity (0.05–0.15 S/cm (80–100 °C; RH 50%–100%)) provided that it is well moistened	(Susai et al., 2001)
Aromatic hydrocarbon-based sulfonated membranes	SPEEK	– Good mechanical durability – Good chemical and thermal stability even at high temperatures	– Decrease in the proton transport coefficient as the water content decreases – Observation of swelling behavior – Conductivity (0.03 S/cm; 100 °C; RH 100%) (0.11 S/cm; 150 °C; RH 100%)	(Kreuer, 2001)
Acid–base polymer complex membranes	PBI/H_3PO_4	– Very stable thermally and dimensionally	– Greater potential for fuel cell at moderate temperatures – Excellent conductivity (0.05 S/cm (180 °C RH 10%) 0.13 S/cm (160 °C) 0.1 S/cm (200 °C)	(Perry et al., 2014)
Inorganic composite membrane	SiO_2/Nafion TiO_2/SPEEK TiO_2/PBI	– Good stability	– Good performance at high temperatures (175 °C, 1000 mW cm^{-2})	(Antonucci et al., 2008) (Lobato et al., 2011)

2.3 Solid (cationic) polymer electrolyte membranes

Grubb, who first described an organic cation exchange membrane as a solid electrolyte for a fuel cell in 1959, used this membrane in an electrochemical cell. Nowadays, it has become one of the most promising systems among all fuel cell systems in terms of its operating style and applications (Rikukawa & Sanui, 2000). In the 1950s, researchers made the first ion exchange membranes, which act as electrolytes for fuel cells. However, it was hard for them to find the best membrane for fuel cell operating conditions (Zaidi & Matsuura, 2008). Since the introduction of the first polymer electrolyte fuel cells in the Gemini space program in the early 1960s, PEMs have undergone a series of technological transformations. Sulfonated polystyrene was used as PEM in PEMFCs introduced by Gemini. However, these solid electrolyte membranes are short-lived due to their poor stability as well as rapid oxidative degradation and are very expensive for commercial applications (Kraytsberg & Ein-Eli, 2014). Due to these problems, the Nafion® ionomer developed by DuPont was replaced by this system shortly after. Replacing the polystyrene sulfonic acid membrane with the Nafion membrane developed by DuPont Chemical is three times more effective in PEMFC performance. A significant increase in oxygen reduction kinetics by decreasing the adsorption of perfluorosulfonic acid anions on the platinum electrocatalyst was provided with these membranes. The proton conductivity was greatly increased due to the presence of more electronegative F atoms than H atoms. CF groups in the structure of the perfluorosulfonic acid polymer are not sensitive to electrochemical oxidation. The stability of perfluorinated sulfonic acid membranes and the lifespan of PEMFCs were significantly increased. Thus, Nafion® has proven to be superior to these membranes in terms of performance and endurance. The development of Nafion® and Flemion®, a perfluorosulfonic acid polymer, has resulted in further progress in this area (Wakizoe et al., 1995; Çelik et al., 2012). Perfluorosulfonated polymer membranes have high proton conductivity and good thermomechanical stability. Because of these various unique features, they have become ideal PEMs over the years. However, these PEMs have several disadvantages, such as low electrode kinetics, a low operating temperature, and a high cost. These drawbacks make it difficult to obtain alternative PEMs. After a while, it was understood that these drawbacks encountered in perfluorosulfonated PEMs could be successfully overcome with PEMs operating at higher temperatures ($>100\,°C$) (Harilal et al., 2020). Thus, many fluorine-free membranes that can be alternatives to Nafion have been studied

(Asensio et al., 2010). By far, many different polymers have been studied in the literature, including polyarylene piperidines (PAP) (Bai et al., 2019), sulfonated polyimides (SPI) (Ye et al., 2006), sulfonated polyether ether ketones (SPEEK) (Kreuer, 2001), and poly (arylene ether sulfone) (PAES) (Li et al., 2017). Comprehensive research has been suggested to develop high-temperature and high-efficiency PEMs using such polymers (Harilal et al., 2020; Rath et al., 2020; Lee et al., 2006). These materials, which are separated according to their operating temperature, stand out in different respects, such as proton conductivity, water uptake, and mechanical and thermal stability (Wong et al., 2019).

2.3.1 PEM Materials

Membrane material development studies play a key role, especially for the commercialization of PEMFCs so that they can be used in daily life. The industry standards for fuel cells accept the following materials as alternatives to Nafion for fuel cell membranes:

- ▷ fluorinated polymer materials including Nafion;
- ▷ sulfonated aromatic hydrocarbon polymer materials;
- ▷ polymer acid complex (PACs) materials;
- ▷ inorganic-organic composite materials.

2.3.1.1 Perfluoro sulfonated membranes

Proton-conducting polymers generally tend to be solid and poorly conductive unless they absorb water. Hydrated polymer electrolytes, whose proton conductivity increases significantly with water content (10^{-2} – 10^{-1} Scm^{-1}), generally contain negatively charged groups in the polymer backbone (Rikukawa & Sanui, 2000).

Perfluorosulfonic acid membranes are low-temperature fuel cell membranes that contain both hydrophilic and hydrophobic phases. In these membranes, which have a multi-phase structure, while the hydrophilic phase serves to retain water, the hydrophobic phase is necessary for the structural integrity of the membrane. Water is very important, as it provides conductivity by helping protons separate from sulfonic acid groups and forms highly mobile hydrated protons. These membranes are hydrated by moisturizing one or both of the reactant gases. The humidification process is essential to maintaining the optimum performance of the membranes (Bose et al., 2011).

PEMFC membranes are actually copolymers made mostly of perfluorocar bon-sulfonic acid ionomers. This copolymer consists of tetrafluoroethylene and diverse perfluorosulfonate ionomers. The best-known of these membranes is Nafion®, produced by DuPont. Other manufacturers, including Asahi Glass (Flemion®), Asahi Chemical (Aciplex®), Chlorine Engineers ("C" membrane), and Dow Chemical, have created membranes similar to Nafion and have sold them commercially or as research and development products (Barbir, 2006). Polyperfluorosulfonic acids with excellent properties such as Nafion®, Aquivion®, Aciplex®, Flemion®, 3Mionomer®, and Hyflon® have become references as PEM materials and have been used in many studies (Souquet-Grumey et al., 2014). The chemical structures of the Nafion membrane and other perfluorinated electrolyte membranes are shown in Figure 2.1.

Almost all of these membranes are characterized as acidic electrolytes in which negative ions are retained in a polymeric matrix. These membranes usually contain a polytetra or trifluoroethylene backbone. These backbones are structures with ion chains ending in sulfonic acid groups. The high thermal and chemical stability and good ionic conductivity of fluorinated polymers depend on the good acidity of the strong tetrafluoroethylene backbone and the presence of sulfonic acid groups in the side chain. Fluorocarbon-based membranes demonstrate good electrical properties in spite of their low water intake. These features are due to the fact that they

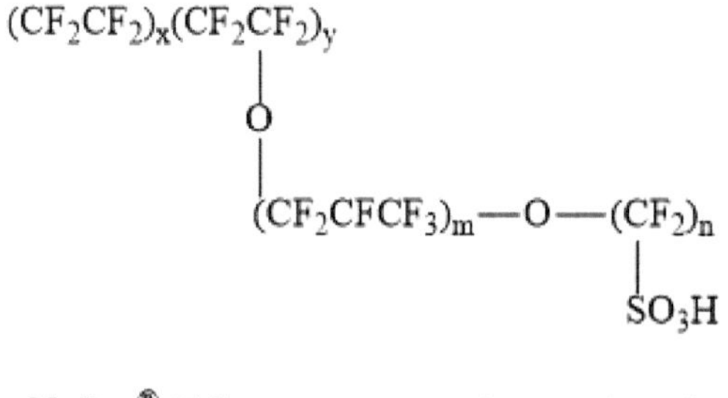

$(CF_2CF_2)_x(CF_2CF_2)_y$

O

$(CF_2CFCF_3)_m$—O—$(CF_2)_n$

SO_3H

Nafion® 117	m>1 or m=1, n=2, x=5-13.5, y=1000
Flemion®	m=0, 1; n=1-5
Aciplex®	m=0, 3; n=2-5, x=1.5-14
Dow membran	m=0, n=2, x=3.6-10

Figure 2.1 Chemical structures of perfluorinated PEMs (Peighambardoust et al., 2010; Ogungbemi et al., 2019; Rikukawa & Sanui, 2000) (Original figure).

have efficient and uniform water channels. Consequently, fluorocarbon-based membranes are accepted as proton conductors in PEMFCs, because these membranes have great electrochemical features and are resistant to harsh conditions (Kim et al., 2015).

Late in the 1960s, DuPont created Nafion®, which continues to be used in low-temperature fuel cells as a state-of-the-art membrane and consists of three parts. These parts consist of the fluorocarbon backbone formed by the repetition of hundreds of $–CF_2–CF–CF_2–$ units similar to the Teflon structure, the $–O–CF_2–CF–O–CF_2–CF_2–$ side chain connecting the fluorocarbon backbone to the last part, and ion groups formed by SO_3^- H^+ ions (Escorihuela et al., 2020; Grujicic & Chittajallu, 2004). Figure 2.2 shows the chemical structure of Nafion. Nafion, a standard and trademarked membrane for PEMFCs today, is one of the most improved polymer materials commercially available (Souzy & Ameduri, 2005; Ogungbemi et al., 2019). Nafion is the archetypal membrane because its proton conductivity is 0.13 S cm^{-1} at 100% relative humidity at 75 ◦C temprature, it is chemically stable, and it has a lifespan of more than 60,000 hours (Devanathan, 2008).

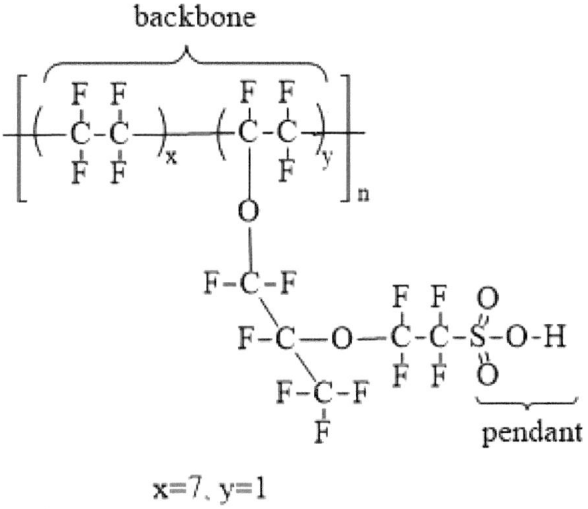

Figure 2.2 Chemical structure of Nafion (Devanathan, 2008) (Original figure).

Since the Nafion® membrane was invented in 1968, there have been big improvements in how long PEMFCs last and how well they work. A lifespan of over 50,000 hours has been obtained with the commercially produced Nafion® 120 and over 10,000 hours with the Dow (Rikukawa & Sanui, 2000; Souzy & Ameduri, 2005). Thinner materials are preferred as membrane materials, as the reduced ionic resistance in PEMFCs increases the membrane-electrode assembly (MEA) performance. The use of thin membrane material also contributes to the reduction of MEA costs (Hogarth & Ralph, 2002). Nafion® membranes in diverse sizes and thicknesses have been produced (Barbir, 2006). The thicknesses of Nafion®-117 and 125 membranes with an equivalent repeat unit molecular weight of 1100 are 175 and 125 µm (micrometer) in dry conditions, respectively. The development of PEMFCs accelerated after Ballard Technologies Corporation reported the possibility of application in electric vehicles using perfluoro membranes developed by Doe Chemical. The Dow membrane, which has a molecular weight of approximately 800 equivalents, is 125-µm (micrometer) thick in the wet state. Developed by Asahi Glass Company, Flemion® R, S, and T are perfluoro membranes with 1000 equivalent molecular weights and 50-, 80-, and 120-µm (micrometer) thicknesses, respectively, in dry conditions. The Aciplex®-S membrane, produced by the Asahi Chemical Industry, produced a series of membranes with a molecular weight of 1000–1200 equivalents and a thickness of 25–100 µm (micrometer) (Rikukawa & Sanui, 2000). The different properties of perfluorinated PEMs are shown in Table 2.2.

Table 2.2 Features of commercial perfluorinated PEMs (Kim et al., 2015; Li et al., 2003; Lee et al., 2006).

Trade name and type of membrane	**Equivalent weight**	**Thickness µm (micrometer)**	**Conductivity (S/cm)**
DuPont Nafion® 120	1200	260	
Nafion® 117	1100	175	0.1
Nafion® 115	1100	125	
Nafion® 112	1100	80	
Asahi Chemicals Aciplex®-S	1000–1200	25–100	–
Asahi Glass Flemion®-T	1000	120	–
Flemion®-S	1000	80	–
Flemion®-R	1000	50	–
Dow Chemical Dow®	800	125	0.2–0.12

Since the middle of the 1980s, the desire to develop PEMFCs has been growing fast. The fuel cells with the best expectations, such as high force, energy density, and long life, were developed. The reduction of activation, mass transfer, and ohmic overpotentials and the improvement of platinum utilization were achieved by positioning the electrocatalysts close to the front surface of the electrodes. Nafion®-117, produced by Dupont Chemical, was used as a proton-conducting membrane in the first stage of studies with low-platinum-loaded electrodes. As a result of these studies, it has been revealed that increasing the power densities depends on reducing the ohmic resistance of the electrolyte. Therefore, it was first tried to produce thinner Nafion® membranes (175 μm ((micrometer)) → 100 μm ((micrometer)) → 50 μm ((micrometer))). Although the results were promising, it was observed that the oxygen electrode was depolarized by the significant transition of hydrogen to the oxygen electrode through the 50-μm (micrometer) electrolyte. In addition, operating with these types of thin membranes at high current densities caused by hot spots leads to the formation of pinholes in the membrane and deterioration of the cell. Therefore, it has been focused on developing alternative membranes. Dow Chemical Company has developed lower equivalent weight experimental membranes and further advanced PEMFC technology (Wakizoe et al., 1995).

Some studies say that the thickness of the membrane and the amount of water in the membrane are the most important factors in how well the PEMFC works in general. So, the goal of recent studies has been to look at how well the membrane works in terms of how well it lets protons and electricity through and how stable it is. Ogungbemi et al. (2019) say that most membranes used in fuel cells, including perfluorinated membranes, are made wet so that the cell can work well at operating temperatures. These membranes have some problems, such as less proton conductivity at low humidity, less stability at medium and high temperatures, a high cost (at least US$ 780/m^2), and the ability for methanol to pass through. The reason for the high cost is the long preparation period and the membrane thickness (Souzy & Ameduri, 2005; Ogungbemi et al., 2019; Tazi & Savadogo, 2000). Other researchers think that if the operating temperature of the membrane is increased above 80 °C and designed without humidification at these temperatures, the cost will decrease, the amount of platinum used will decrease, and there will be a significant increase in fuel cell performance (Ogungbemi et al., 2019).

The purpose of alternative membrane material development studies is to obtain low-cost membranes that can operate at high temperatures and relative humidity with proton conductivity close to the Nafion membrane. These studies aim to develop membranes that can retain water at high temperatures or show good conductivity without water. Among the strategies included in the research are modifying existing membranes, adding hygroscopic oxides and heteropoly acids to the structure to give the membrane water retention, using polymers with aromatic backbones, or using a less volatile solvent instead of water as a proton source (Devanathan, 2008).

2.3.1.2 Aromatic hydrocarbon-based sulfonated membranes

Chandan et al. (2013) found that sulfonated hydrocarbon polymers are a good type of membrane for PEMFCs that work at higher temperatures. People tend to choose polyaryl membranes made from aromatic hydrocarbons, especially polyether ketones, because they are cheaper and more stable. Instead of Nafion, these membranes can work at high temperatures. They are a promising alternative because they are cheap, stable at high temperatures, high pressures, and in the presence of oxygen, and they can be processed in a wide range of chemistries (Kreuer, 2001; Devanathan, 2008). Aromatic structures are added right to the polymer backbone to make aromatic hydrocarbon-based membranes more stable and give them better properties. Although aromatic rings offer thermal and mechanical stability, the desired flexibility and oxidative stability in these membranes are achieved by adding ether bonds between the benzene rings. The benzene rings linked together without ether bonds result in solid polymers unsuitable for PEMs (Devanathan, 2008; Kim et al., 2015). The biggest advantage of hydrocarbon-based membranes is that the polymeric structure can be easily designed to have desired properties. Thus, hydrocarbon membranes with very good properties can be prepared by using different types of monomers in order to control the reaction conditions. The fact that these monomers used in the general, the carbon backbone of hydrocarbon polymers containing polar groups provides significant water uptake over an expansive temperature range. In addition, these polymers advantages. In general, the carbon backbone of hydrocarbon polymers containing polar groups provides significant water uptake over an expansive temperature range. In addition, these polymers are easy to recycle using traditional methods (Rikukawa & Sanui, 2000; Kim et al., 2015).

Since hydrocarbon polymers have various advantages and the operating temperatures of fluorinated membranes are limited, high temperature, chemically and thermally more stable polymers such as poly(ether ketones) (PEK) and their derivatives (SPEEK, etc.) are an alternative (Kreuer, 2001). Polystyrene sulfonate (PSS) (Chen et al., 2004), sulfonated polysulfones

(SPSU) (Lufrano et al., 2000), sulfonated poly (4-phenoxybenzoyl-1,4-phenylene) (SPPBP) (Yoshida-Hirahara et al., 2020), poly (phenylene sulfide) (SPPS) (Li et al., 2017), and polybenzimidazole (PBI) (Wainright et al., 1995; Staiti et al., 2001) are used as proton conductive materials (Neburchilov et al., 2007; Rikukawa & Sanui, 2000). Chemical structures of polymer electrolyte membranes based on aromatic hydrocarbon polymers are shown in Figure 2.3.

As an alternative to expensive perfluorosulfonic acid membranes like Nafion, the best membranes are sulfonated polyarylenetherketone membranes like sulfonated polyetheretherketone (SPEEK), which do not contain fluorine. Although the fluorine backbone in Nafion-type membranes helps to

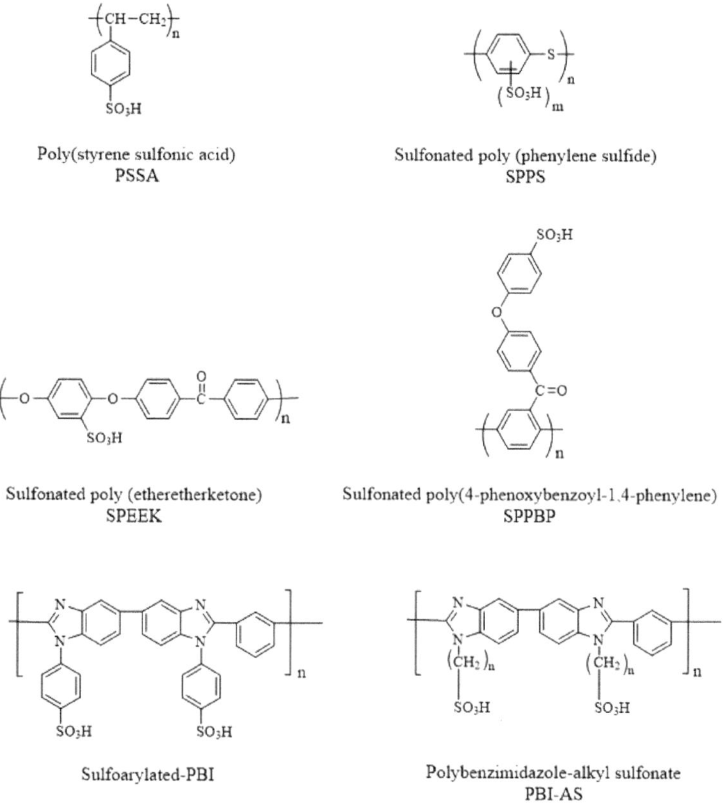

Poly(styrene sulfonic acid)
PSSA

Sulfonated poly (phenylene sulfide)
SPPS

Sulfonated poly (etheretherketone)
SPEEK

Sulfonated poly(4-phenoxybenzoyl-1,4-phenylene)
SPPBP

Sulfoarylated-PBI

Polybenzimidazole-alkyl sulfonate
PBI-AS

Figure 2.3 Chemical structures of some of these aromatic polymers (Rikukawa & Sanui, 2000; Peighambardoust et al., 2010; Li et al., 2003; Ogungbemi et al., 2019; Lee et al., 2006) (Original figure).

increase proton conductivity and acidity of the sulfonic acid group, the cost, water content, and methanol permeability of non-fluorine sulfonated polyetheretherketone membranes are lower. By examining the differences in the microstructure and acidity of these systems, it may be possible to explain the differences in the behavior of these two membrane types (Fontananova et al., 2010; Iulianelli & Basile, 2012).

In perfluorosulfonic acid polymers, the perfluorinated backbone is very water-repellent, while the sulfonic acid groups at the ends are very water-loving. When PFSA membranes are moistened, only the hydrophilic portions are hydrated to maintain conductivity. Hydrophobic parts, on the other hand, provide mechanical strength. Therefore, these membranes have a high water uptake, but a high sensitivity to relative humidity. In sulfonated hydrocarbon polymers, the hydrophobicity of the polymer backbone is lower. Besides, it has lower acidity (Nafion pKa: −6, SPEEK pKa: −1) and polarity compared to Nafion (Li et al., 2003). The absence of side chains and rigid backbone structure in the SPEEK membrane causes different transport properties and morphological stability (Kreuer, 2001; Devanathan, 2008). Polymers with side chains between the polymer backbone and the sulfonic acid group are expected to have better stability to hydrolysis than polymers with sulfonic acid groups directly attached to the polymer backbone. These flexible side chains, which provide nanophase division of hydrophilic and hydrophobic areas, improve proton conductivity and provide dimensional stability to the membrane (Yoshida-Hirahara et al., 2020). The presence of a good hydrophilic domain ensures the carriage of protons and water, while a good hydrophobic part ensures the morphological stability and insolubility of the polymer in water (Iulianelli & Basile, 2012).

With the small angle X-ray scattering (SAXS) tests, it was found that the separation between hydrophilic and hydrophobic areas was not as clear in SPEEK polymers. This is because the hydrophobic backbone is smaller and the sulfonic acid group has less acidity and less polarity. The direct attachment of sulfonic acid groups to the rigid backbone in the SPEEK membrane prevents a clear microphase separation from the microphase. Consequently, the SPEEK polymer microstructure has less flexibility and greater resistance to water and methanol transport compared to Nafion (Fontananova et al., 2010). The water-filled channels of the sulfonated polyetheretherketone membranes are narrower than the Nafion membrane, and they are also more branched, with less separated and more dead-end channels (Figure 2.4). This means that the hydrophilic/hydrophobic interface is larger, and thus the average separation of adjacent sulfonic acid groups is greater. Stronger

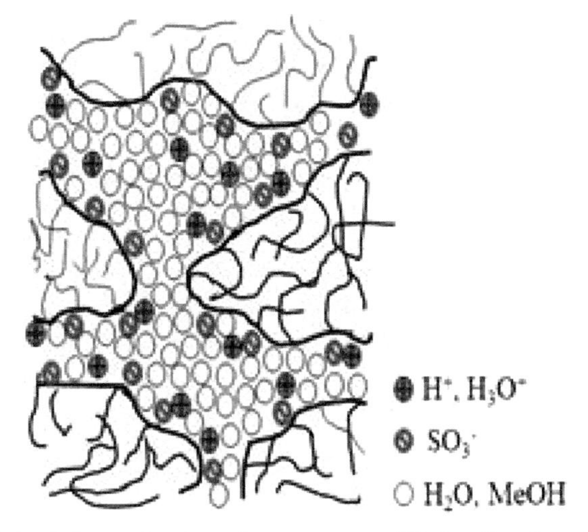

Figure 2.4 Schema showing microstructures of Nafion (Junoh et al., 2020).

retention of water in aromatic polymers with narrow channels causes the dielectric constant of water during hydration to decrease significantly. All these differences cause a decrease in the proton conductivity of the SPEEK membrane. In order for it to have a proton conductivity close to Nafion, it must be more humidified and have a higher ion exchange capacity. The high ion exchange capacity of the SPEEK membrane leads to increased water uptake, excessive swelling, and weakening of the mechanical properties. To overcome these problems, the SPEEK membrane can be blended or cross-linked with different polymers (Kreuer, 2001; Iulianelli & Basile, 2012; Devanathan, 2008).

2.3.1.3 Acid–base polymer complex membranes

It is seen that the morphological stability of acidic polymers obtained by covalent cross-linking or acid/base complexes reduces swelling, water absorption, and methanol transmission. However, in their dry state, new troubles such as low conductivity and brittleness arise. Differently, approaches based on complexes of basic polymers with oxo acids result in conductivity close to that of pure acid (Kreuer, 2001).

Acid–base complexes, which are considered a suitable alternative for fuel cell membranes, are membranes that can maintain their good conductivity at high temperatures because they are not damaged by the effects of dehydration. Proton conduction in these membranes is achieved by combining an acid component with an alkaline polymer (Peighambardoust et al., 2010; Smitha et al., 2005). The structures of some remarkable acidic and basic polymers are shown in Figure 2.5.

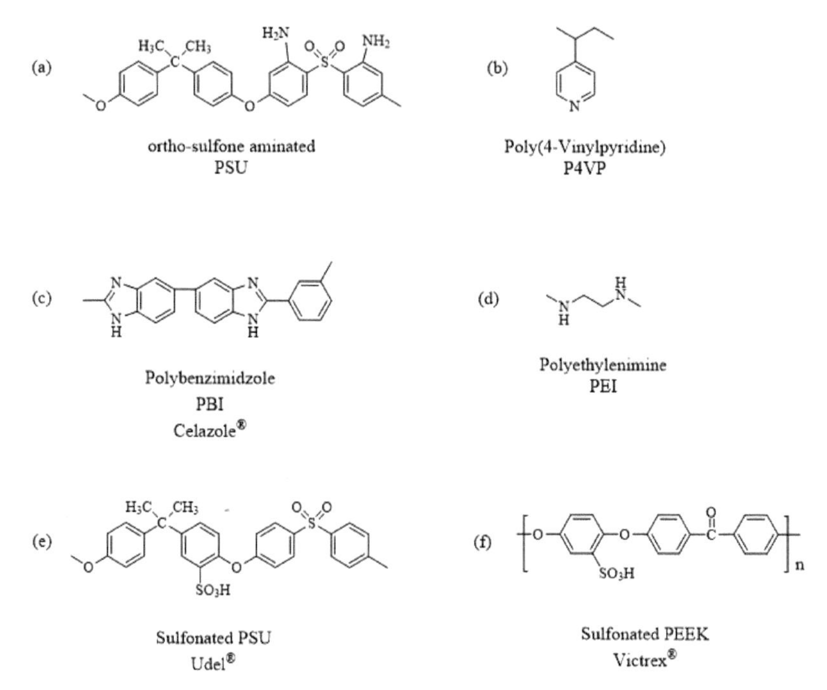

Figure 2.5 Chemical structure of (a–d) basic polymers and(e–f) acidic polymers (Ogungbemi et al., 2019; Peighambardoust et al., 2010; Smitha et al., 2005; Kerres et al., 1999) (Original figure).

Li et al. (2003) found that basic polymers can be charged with an amphoteric acid that lets protons move around and acts as both a proton acceptor and a proton donor. Strong acids like sulfuric acid (H_2SO_4) or phosphoric acid (H_3PO_4) react with ether, alcohol, amide, imine, or imide groups that have basic parts in their structures. Because of how these polymers are made, they can form hydrogen bonds with acids. That is, basic polymers help the acid break down by acting as a kind of solvent. Even the anhydrous forms (100%) of H_3PO_4 and H_2SO_4 have an effective proton conductivity. Because these acids have an unparalleled proton conduction mechanism with their self-ionization and spontaneous dehydration properties, various basic polymers have been investigated to prepare complex acid–base electrolytes. Complexes are obtained from many basic polymers such as polyacrylamide

(PAAM), polyethylene oxide (PEO), polyethyleneimine (PEI), polybenzimidazoles (PBI), and polyvinyl alcohol (PVA), and strong acids such as H_2SO_4 or H_3PO_4 have been determined to have good proton conductivities in both hydrated and non-hydrated conditions (Li et al., 2003; Lee et al., 2006). The structure of these acid–base complex membranes is shown in Figure 2.6.

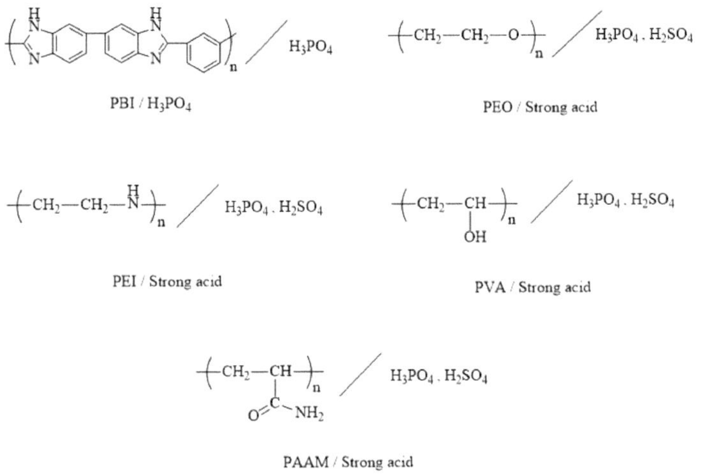

PBI / H_3PO_4

PEO / Strong acid

PEI / Strong acid

PVA / Strong acid

PAAM / Strong acid

Figure 2.6 Chemical structure of acid/base complex polymer electrolytes (Lee et al., 2006; Rikukawa & Sanui, 2000) (Original figure).

At temperatures above 100 °C, the conductivity of the membranes increases with the increase in acid content. However, this leads to a decrease in mechanical durability. Efforts to improve mechanical strength have been directed toward cross-linking polymers or adding different inorganic fillers to polymers (Li et al., 2003).

2.3.1.4 Inorganic composite membranes

Composite membranes produced by doping polymers with a filler material are of great interest. The main purpose of adding fillers to polymers, which is a way to increase the temperature tolerance of classic PEM materials, is to increase conductivity at high temperatures and low humidity and to improve water uptake and hold. Doping of various inorganic fillers into many polymer electrolyte types has been successfully carried out. These materials are inorganic filling materials such as hygroscopic oxides (SiO_2, TiO_2, ZrO_2, and Al_2O_3), clays (Montmorillonite), zeolites, mineral acids (HCl and H_3PO_4), heteropoly acids, and zirconium phosphates (ZrP) (Chandan et al., 2013).

The goal of making composite membranes is to improve their mechanical strength, thermal stability, ability to pass protons, and many other properties. For example, when the degree of sulfonation of sulfonated polymers is increased to make the proton conductivity of the polymers higher, the membrane swells, which makes it less strong. Li et al. (2003) say that this mechanical behavior is fixed by adding an inorganic part to the polymer.

Li et al. (2003) and Ogungbemi et al. (2019) say that inorganic/organic composites are made by fusing an acid component with an alkaline polymer so that they can work at high temperatures and have more conductivity. Nowadays, composite materials fused to polymers are increasingly used due to their unique features. Some of these features are a high durability-to-weight ratio and thermal stability, low thermal dilation, and very good corrosion and chemical resistance (Abdolmaleki & Molavian, 2015). These membranes are inexpensive and recyclable, with a high operating temperature range and good water uptake compared to other membrane kinds (Ogungbemi et al., 2019). In addition to these, the reason why these membranes are of interest is that there are significant changes in their mechanical, thermal, electrical, and magnetic properties compared to the pure states of the organic and inorganic components that make up these membranes (Xu, 2005).

The effects of different inorganic fillers on different polymer matrixes are also quite different. The advantages and disadvantages of composite membranes obtained from these inorganic fillers and polymer matrices are shown in Table 2.3.

Table 2.3 Advantages and disadvantages of the different modifications on different PEMs (references in the table are valid).

Type of filler material	PEM type	Advantages and disadvantages	Ref.
Inorganic oxides, MO_2 (SiO_2, TiO_2, ZrO_2, Al_2O_3)	Nafion, PBI	– High water uptake – Low water swelling – Increase in performance – High cost	(Antonucci et al., 2008) (Lee et al., 2020)
Conductive Polymers (Polyaniline (PANi) Polypyrrole (PPy))	Nafion, SPEEK	– High selectivity – Low proton conductivity – Low water uptake and swelling – High cost	(Yang et al., 2009) (Li et al., 2006)
Proton conductive fillers (Heteropolyacid (HPA), $Cs_{2.5}H_{0.5}PW_{12}O_{40}$, tungstophosphoric, phosphotungestic acid)	Nafion, SPEEK	– High proton conductivity – Self-humidifying properties – Good fuel cell performance in high temperature – High leaching rate – High cost	(Ramani et al., 2004) (Zhang et al., 2008)

Table 2.3 *Continued*

Type of filler material	PEM type	Advantages and disadvantages	Ref.
Acid–base complexes (PBI/H_3PO_4, $Nafion/PBI/H_3PO_4$)	Nafion, PBI	– Good mechanical strength at high temperature – Low gas permeability – Perfect thermo-chemical stability – High proton conductivity – High cost	(Perry et al., 2014) (Zhai et al., 2007)

The additive material used for the performance of the materials in inorganic composite membranes and the preparation of these additives are very important (Chandan et al., 2013). However, there are several important troubles in the development of composite membranes where optimal conditions are not created.

- to determine the most suitable inorganic filler and dispersion circumstances for the chosen polymer matrix;
- setting optimum conditions for the concentration of doped superior composite PEMs;
- to determine the most suitable synthesis conditions and membrane casting conditions for composite membranes;
- to have knowledge about the interactivities between the polymer matrix, filler, and/or specified solvent (Tripathi & Shahi, 2011).

2.4 Conductivity Mechanism

Proton transfer in PEMs proceeds through two mechanisms, namely hopping (Grotthuss) and vehicle, depending on the hydration level of the membrane. In the Grotthuss mechanism, proton transfer occurs when protons sequentially jump from one place to another via H-bonds along the chain. This mechanism, suggested by Theodore von Grotthuss in 1806, aims to explain how proton transmission occurs between water molecules. Proton conduction by the Grotthus mechanism is shown in Figure 2.7. In the vehicle mechanism, proton transfer occurs by diffusion and migration of a vehicle or solvent containing protons such as H_3O^+, $H_5O_2{}^+$ (Zundel cation), and $H_9O_4{}^+$ (Eigen cation). This mechanism, in which the motion of protons is usually assisted by molecules associated with doped acidic groups, was suggested by Kreuer, Rabenau, and Weppner in 1982. Proton conduction by the vehicle mechanism is shown in Figure 2.8. In polymeric membranes, where two mechanisms can

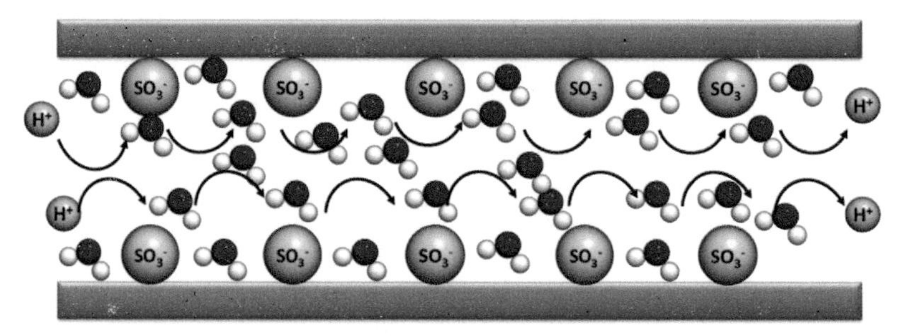

Figure 2.7 Schematic of the Grotthuss-type transport mechanism (Escorihuela et al., 2020)..

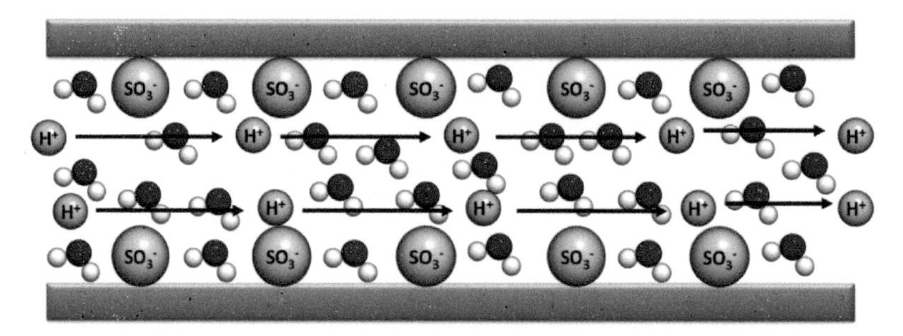

Figure 2.8 Schematic of vehicle-type transport mechanism (Escorihuela et al., 2020).

be observed depending on the diffusion of protons, the vehicle mechanism significantly contributes to the increase in conductivity. Although it is advantageous to use water as a low-cost, environmental-friendly, and fast proton transporter as a solvent in proton conduction, it becomes a problem to dry out as it rises to high temperatures (Escorihuela et al., 2020; Wong et al., 2019; Yan & Xie, 2012; Zuo et al., 2012).

In the Grotthuss-type mechanism, proton transmission occurs relatively faster than the vehicle mechanism, because there is a short distance and sequential transition between the hydronium (H_3O^+) ions in the environment and the neighboring acid groups attached to the membranes. In acid–base complex membranes, which are frequently used in recent studies, acid components are the source of protons, while base components help proton transfer by the Grotthuss mechanism (Figure 2.9) (Escorihuela et al., 2020; Zuo et al., 2012).

The water bridge formed by the free water molecules in the perfluorinated (e.g., Nafion) and aromatic hydrocarbon based sulfonated (e.g., SPEEK) membranes and the sulfonic acid groups attached to these membranes are liable for proton transfer. Increasing the temperature and decreasing the

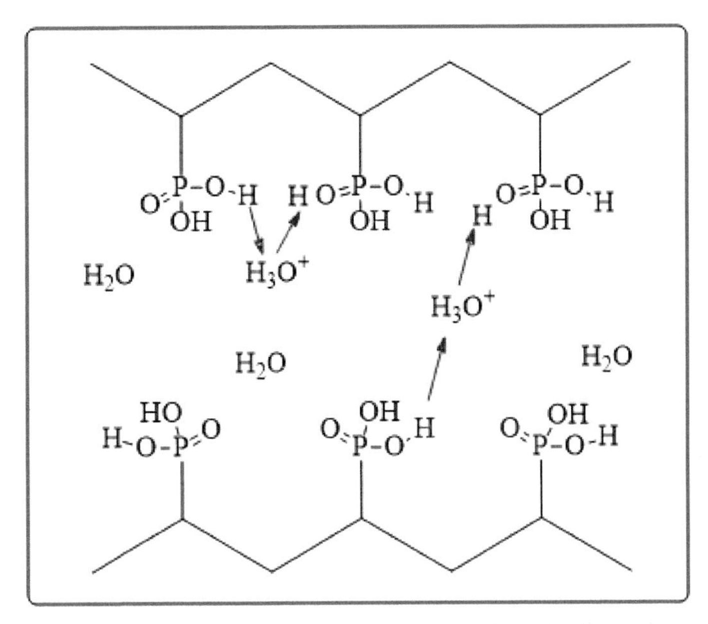

Figure 2.9 Diagram of the Grotthuss-type transport mechanism of acid complex membranes (Wong et al., 2019) (Original figure).

relative humidity cause a disconnection between the ion channels. As a result, since the operating temperature is affected by the water content of the membrane, the proton transport mechanism and speed are also affected by this situation (Wong et al., 2019).

The proton transfer mechanism in composite membranes is more complex because conductivity covers both the surface and chemical features of the organic and inorganic parts of these membranes (Escorihuela et al., 2020). The addition of a hygroscopic additive to a PEM increases the degree of swelling of the membrane at low relative humidity and prevents fuel passage. This facilitates proton transport. In addition, conductive materials prevent the molecular emigration of undesirable species to the membrane by tightening the pores of the polymer (Bose et al., 2011).

2.5 Membrane Preparation

Since the cost of polymeric materials and components that make up the membrane also affects the total production cost, more effort is needed to reduce the cost. Therefore, the need to develop cheaper and higher-performance materials is becoming more and more important. Membranes are prepared or fabricated with different methods selected according to the membrane type, existing materials, and equipment. There are many methods used to prepare or fabricate membranes. These methods differ from each other according to different production techniques. Some of these methods are the plasma method, irradiation grafting method, phase inversion method, cross-linking method, sol–gel method, direct copolymerization, covered chemical polymerization, solution casting method, layer-by-layer self-assembly technique, ultrasonic coating technique, ultraviolet polymerization, *in situ* reduction, decal transfer method, and catalyst coated membrane method. It would be beneficial to use these methods in PEM synthesis according to their applicability in the fuel cell, cost-effectiveness, and ease of production technique (Ogungbemi et al., 2019; Mohammad & Asiri, 2017; Wong et al., 2019).

Some of the most commonly used membrane preparation or fabrication methods in the literature for fuel cells, with their pros and cons, are as follows.

Cross-linking method: This method, which improves membrane properties, is prepared by combining two compounds with cross-links. It has been stated that these structures solve the swelling problem and increase the tensile strength. In addition, studies have reported that it is recyclable, and its mechanical and thermal properties up to 200 °C are better than Nafion. However, in addition to its disadvantages, such as reduced water uptake because of cross-linking and decreased activation energy, it needs further development (Ogungbemi et al., 2019).

Sol–gel method: This method, which is mostly used to prepare organic–inorganic nanocomposite membranes, is environmental-friendly because it is prepared at low temperatures. It is prepared by combining two important materials, such as sol and gel, an organic polymer matrix, and a pure inorganic part. Although it has installation flexibility and versatility, it is a method in which physical and chemical processes are complex (Mohammad & Asiri, 2017; Ogungbemi et al., 2019).

Direct polymerization method: This method, which is often used to make membranes, is made by combining different monomers into a bigger molecule. In addition, this method, which has very high proton conductivity and fuel cell performance, is versatile and creative in terms of developing many new products. However, the properties of the monomers to be used in polymerization in this method should be well investigated because the effect of the polymer obtained with the monomers used on the membrane is quite high (Ogungbemi et al., 2019).

Solution casting method: It is the method by which the membrane is obtained by evaporation of the solvent after the polymer is dissolved with the relevant solvent before casting it into glass petri dishes. With this method, usually an asymmetrical structure, intensive, rough, thin skin, and porous support layer membranes are obtained (Wong et al., 2019).

2.6 Characterization of Polymers and Membranes

There are different ways to characterize polymers and membranes in order to find out what their properties are. By using these techniques, physical and chemical properties are determined.

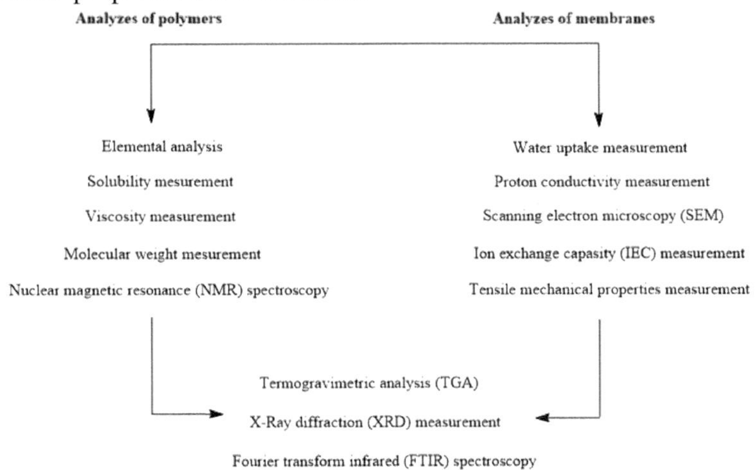

2.7 Conclusions

There are many physical and chemical parameters that affect the performance of a fuel cell, such as temperature, pressure, relative humidity, channel length, channel width, channel height, surface area, thickness, porosity, permeability, concentration or density, mass and water transport, electrical or ionic conductivity, and thermal conductivity (Singh et al., 2022; Ogungbemi et al., 2021). Therefore, each fuel cell component must be designed in accordance with these parameters. One of the most important components that affect fuel cell performance is the type of PEM used. Because many simultaneous reactions take place at the interface of the PEM while the fuel cell is operating, the smallest change that may occur in one of the parameters causes significant changes in other parameters. Therefore, PEMs have an overall effect on fuel cells (Ogungbemi et al., 2019). PEMs used in fuel cells have many tasks, such as providing and facilitating proton conduction, creating a barrier between the anode and the cathode, and preventing electron transfer. The most common PEM still used in academic and commercial applications is Nafion. However, there are limitations to the use of Nafion membrane, such as the need for an additional humidification system, loss of performance at high temperatures, and high cost. Therefore, it is aimed to develop alternative PEMs that have the potential to replace Nafion with excellent physical properties, thermal and chemical stability, mechanical strength, conductivity, and low cost at high temperatures and low relative humidity or non-humid conditions. However, developing new chemical groups with a much better proton conductivity than sulfonic acid groups and making these new groups applicable to polymeric PEMs, that is, discovering alternative new PEMs, is innovative but costly and challenging (Zhang et al., 2012). As a result, there are many studies in the literature aiming to develop alternative PEMs with improved properties, including chemical modifications of polymers. However, the efforts of researchers to develop alternative PEMs with optimal properties still continue. In this chapter, many PEMs with different types and properties developed for application in PEMFCs are examined and discussed in various aspects.

References

[1] A. Perry, K., L. More, K., Andrew Payzant, E., Meisner, R. A., Sumpter, B. G., & Benicewicz, B. C. (2014). A comparative study of phosphoric acid-doped m-PBI membranes. *Journal of Polymer Science Part B: Polymer Physics*, 52(1), 26-35. DOI: 10.1002/polb.23403

[2] Abdolmaleki, A., & Molavian, M. R. (2015). Synthesis and Characterization of Co Nanocomposite Based on Poly (benzimidazole-amide) Matrix and Their Behavior as Catalyst in Oxidation Reaction. *Polymer-Plastics Technology and Engineering*, 54(12), 1241-1250. DOI: 10.1080/03602559.2015.1010214

[3] Antonucci, V., Di Blasi, A., Baglio, V., Ornelas, R., Matteucci, F., Ledesma-Garcia, J., ... & Aricò, A. S. (2008). High temperature operation of a composite membrane-based solid polymer electrolyte water electrolyser. *Electrochimica Acta*, 53(24), 7350-7356. DOI: 10.1016/j.electacta.2008.04.009

[4] Asensio, J. A., Sánchez, E. M., & Gómez-Romero, P. (2010). Proton-conducting membranes based on benzimidazole polymers for

high-temperature PEM fuel cells. A chemical quest. *Chemical Society Reviews*, 39(8), 3210-3239. DOI: 10.1039/b922650h

[5] Atifi, A., Mounir, H., & El Marjani, A. (2014, October). Effect of internal current, fuel crossover, and membrane thickness on a PEMFC performance. *In 2014 International Renewable and Sustainable Energy Conference* (IRSEC) (pp. 907-912). IEEE. DOI: 10.1109/IRSEC.2014.7059860

[6] Bai, H., Peng, H., Xiang, Y., Zhang, J., Wang, H., Lu, S., & Zhuang, L. (2019). Poly (arylene piperidine) s with phosphoric acid doping as high temperature polymer electrolyte membrane for durable, high-performance fuel cells. *Journal of Power Sources*, 443, 227219. DOI: 10.1016/j.jpowsour.2019.227219

[7] Barbir, F. (2006). *PEM fuel cells. In Fuel Cell Technology*. London: Springer, 27-51, DOI: 10.1007/1-84628-207-1_2

[8] Bose, S., Kuila, T., Nguyen, T. X. H., Kim, N. H., Lau, K. T., & Lee, J. H. (2011). Polymer membranes for high temperature proton exchange membrane fuel cell: recent advances and challenges. *Progress in Polymer Science*, 36(6), 813-843. DOI: 10.1016/j.progpolymsci.2011.01.003

[9] Brandon, N. P., Skinner, S., & Steele, B. C. (2003). Recent advances in materials for fuel cells. *Annual Review of Materials Research*, 33(1), 183-213. DOI: 10.1146/annurev.matsci.33.022802.094122

[10] Cao, Y. C., Xu, C., Wu, X., Wang, X., Xing, L., & Scott, K. (2011). A poly (ethylene oxide)/graphene oxide electrolyte membrane for low temperature polymer fuel cells. *Journal of Power Sources*, 196(20), 8377-8382. DOI: 10.1016/j.jpowsour.2011.06.074

[11] Chandan, A., Hattenberger, M., El-Kharouf, A., Du, S., Dhir, A., Self, V., ... & Bujalski, W. (2013). High temperature (HT) polymer electrolyte membrane fuel cells (PEMFC)–A review. *Journal of Power Sources*, 231, 264-278. DOI: 10.1016/j.jpowsour.2012.11.126

[12] Chen, F., Su, Y. G., Sọong, C. Y., Yan, W. M., & Chu, H. S. (2004). Transient behavior of water transport in the membrane of a PEM fuel cell. *Journal of Electroanalytical Chemistry,* 566(1), 85-93. DOI: 10.1016/j.jelechem.2003.11.016

[13] Chen, H., Wang, S., Liu, F., Wang, D., Li, J., Mao, T., ... & Wang, Z. (2020). Base-acid doped polybenzimidazole with high phosphoric acid retention for HT-PEMFC applications. *Journal of Membrane Science*, 596, 117722. DOI: 10.1016/j.memsci.2019.117722

[14] Chen, S. L., Krishnan, L., Srinivasan, S., Benziger, J., & Bocarsly, A. B. (2004). Ion exchange resin/polystyrene sulfonate composite membranes for PEM fuel cells. *Journal of Membrane Science*, 243(1-2), 327-333. DOI: 10.1016/j.memsci.2004.06.037

[15] Çelik, S. Ü., Bozkurt, A., & Hosseini, S. S. (2012). Alternatives toward proton conductive anhydrous membranes for fuel cells: Heterocyclic protogenic solvents comprising polymer electrolytes. *Progress in Polymer Science*, 37(9), 1265-1291. DOI: 10.1016/j.progpolymsci.2011.11.006

[16] Devanathan, R. (2008). Recent developments in proton exchange membranes for fuel cells. *Energy & Environmental Science*, 1(1), 101-119. DOI: 10.1039/B808149M

[17] Ergün, D., (2009). Hıgh temperature proton exchange membrane fuel cells, *MS Thesis*, METU, Ankara.

[18] Escorihuela, J., Olvera-Mancilla, J., Alexandrova, L., Del Castillo, L. F., & Compañ, V. (2020). Recent progress in the development of composite membranes based on polybenzimidazole for high temperature proton exchange membrane (PEM) fuel cell applications. *Polymers*, 12(9), 1861. DOI: 10.3390/polym12091861

[19] Fiuza, R. A., Santos, I. V., Fiuza, R. P., José, N. M., & Boaventura, J. S. (2011, January). Characterization of electrolyte polyester membranes for application in PEM fuel cells. *In Macromolecular Symposia* (Vol. 299, No. 1, pp. 234-240). Weinheim: WILEY-VCH Verlag. DOI: 10.1002/masy.200900142

[20] Fontananova, E., Trotta, F., Jansen, J. C., & Drioli, E. (2010). Preparation and characterization of new non-fluorinated polymeric and composite membranes for PEMFCs. *Journal of Membrane Science*, 348(1-2), 326-336. DOI: 10.1016/j.memsci.2009.11.020

[21] Grujicic, M., & Chittajallu, K. M. (2004). Design and optimization of polymer electrolyte membrane (PEM) fuel cells. *Applied Surface Science*, 227(1-4), 56-72. DOI: 10.1016/j.apsusc.2003.10.035

[22] Guo, Q., Pintauro, P. N., Tang, H., & O'Connor, S. (1999). Sulfonated and crosslinked polyphosphazene-based proton-exchange membranes. *Journal of Membrane Science*, 154(2), 175-181. DOI: 10.1016/S0376-7388(98)00282-8

[23] Guo, Z., Xu, X., Xiang, Y., & Lu, S. (2015). New anhydrous proton exchange membranes for high-temperature fuel cells based on PVDF–PVP blended polymers. *Journal of Materials Chemistry A*, 3(1), 148-155. DOI: 10.1039/C4TA04952G

[24] Haider, R., Wen, Y., Ma, Z. F., Wilkinson, D. P., Zhang, L., Yuan, X., ... & Zhang, J. (2021). High temperature proton exchange membrane fuel cells: progress in advanced materials and key technologies. *Chemical Society Reviews*, 50(2), 1138-1187. DOI: 10.1039/D0CS00296H

[25] Harilal, Nayak, R., Ghosh, P. C., & Jana, T. (2020). Cross-Linked Polybenzimidazole Membrane for PEM Fuel Cells. *ACS Applied Polymer Materials*, 2(8), 3161-3170. DOI: 10.1021/acsapm.0c00350

[26] Hasiotis, C., Qingfeng, L., Deimede, V., Kallitsis, J. K., Kontoyannis, C. G., & Bjerrum, N. J. (2001) Development and characterization of acid-doped polybenzimidazole/sulfonated polysulfone blend polymer electrolytes for fuel cells. *Journal of the Eelectrochemical Society*, 148(5), A513.

[27] Herring, A. M. (2006). Inorganic–polymer composite membranes for proton exchange membrane fuel cells. Journal of Macromolecular Science, Part C: *Polymer Reviews*, 46(3), 245-296. DOI: 10.1080/00222340600796322

[28] Hogarth, M. P., & Ralph, T. R. (2002). Catalysis for low temperature fuel cells. *Platinum Metals Review*, 46(4), 146-164.

[29] Iulianelli, A., & Basile, A. (2012). Sulfonated PEEK-based polymers in PEMFC and DMFC applications: A review. *International Journal of Hydrogen Energy*, 37(20), 15241-15255. DOI: 10.1016/j.ijhydene.2012.07.063

[30] Jung, G. B., Tseng, C. C., Yeh, C. C., & Lin, C. Y. (2012). Membrane electrode assemblies doped with H3PO4 for high temperature proton exchange membrane fuel cells. *International Journal of Hydrogen Energy*, 37(18), 13645-13651. DOI: 10.1016/j.ijhydene.2012.02.054

[31] Junoh, H., Jaafar, J., Nordin, N. A. H. M., Ismail, A. F., Othman, M. H. D., Rahman, M. A., ... & Yusof, N. (2020). Performance of polymer electrolyte membrane for direct methanol fuel cell application: Perspective on morphological structure. *Membranes*, 10(3), 34. DOI: 10.3390/membranes10030034

[32] Kerres, J., Ullrich, A., Meier, F., & Häring, T. (1999). Synthesis and characterization of novel acid–base polymer blends for application in membrane fuel cells. *Solid State Ionics*, 125(1-4), 243-249. DOI: 10.1016/S0167-2738(99)00181-2

[33] Kim, D. J., Jo, M. J., & Nam, S. Y. (2015). A review of polymer–nanocomposite electrolyte membranes for fuel cell application. *Journal of Industrial and Engineering Chemistry*, 21, 36-52. DOI: 10.1016/j.jiec.2014.04.030

[34] Kraytsberg, A., & Ein-Eli, Y. (2014). Review of advanced materials for proton exchange membrane fuel cells. *Energy & Fuels*, 28(12), 7303-7330. DOI: 10.1021/ef501977k

[35] Kreuer, K. D. (2001). On the development of proton conducting polymer membranes for hydrogen and methanol fuel cells. *Journal of Membrane Science*, 185(1), 29-39. DOI: 10.1016/S0376-7388(00)00632-3

[36] Kurz, T., Küfner, F., & Gerteisen, D. (2018). Heating of Low and High Temperature PEM Fuel Cells with Alternating Current. *Fuel Cells*, 18(3), 326-334. DOI: 10.1002/fuce.201700145

[37] Kusoglu, A., Karlsson, A. M., Santare, M. H., Cleghorn, S., & Johnson, W. B. (2006). Mechanical response of fuel cell membranes subjected to a hygro-thermal cycle. *Journal of Power Sources*, 161(2), 987-996. DOI: 10.1016/j.jpowsour.2006.05.020

[38] Lee, J., Jung, J., Han, J. Y., Kim, H. J., Jang, J. H., Lee, H. J., ... & Nam, S. Y. (2014). Effect of membrane electrode assembly fabrication method on the single cell performances of polybenzimidazole-based high temperature polymer electrolyte membrane fuel cells. *Macromolecular Research*, 22(11), 1214-1220. DOI: 10.1007/s13233-014-2167-x

[39] Lee, J. S., Quan, N. D., Hwang, J. M., Lee, S. D., Kim, H., Lee, H., & Kim, H. S. (2006). Polymer electrolyte membranes for fuel cells. *Journal of Industrial and Engineering Chemistry*-Seoul-, 12(2), 175.

[40] Lee, S., Seo, K., Ghorpade, R. V., Nam, K. H., & Han, H. (2020). High temperature anhydrous proton exchange membranes based on chemically-functionalized titanium/polybenzimidazole composites for fuel cells. *Materials Letters*, 263, 127167. DOI: 10.1016/j.matlet.2019.127167

[41] Li, Q., He, R., Jensen, J. O., & Bjerrum, N. J. (2003). Approaches and recent development of polymer electrolyte membranes for fuel cells operating above 100 C. *Chemistry of Materials*, 15(26), 4896-4915. DOI: 10.1021/cm0310519

[42] Li, Q., Jensen, J. O., Savinell, R. F., & Bjerrum, N. J. (2009). High temperature proton exchange membranes based on polybenzimidazoles for fuel cells. *Progress in Polymer Science*, 34(5), 449-477. DOI: 10.1016/j.progpolymsci.2008.12.003

[43] Li, X., Wang, S., Zhang, H., Lin, C., Xie, X., Hu, C., & Tian, R. (2021). Sulfonated poly (arylene ether sulfone) s membranes with distinct microphase-separated morphology for PEMFCs. *International Journal of Hydrogen Energy*, 46(68), 33978-33990. DOI: 10.1016/j.ijhydene.2021.07.199

[44] Li, X., Liu, C., Xu, D., Zhao, C., Wang, Z., Zhang, G., ... & Xing, W. (2006). Preparation and properties of sulfonated poly (ether ether ketone) s (SPEEK)/polypyrrole composite membranes for direct methanol fuel cells. *Journal of Power Sources*, 162(1), 1-8. DOI: 10.1016/j.jpowsour.2006.06.030

[45] Li, Y., Zhang, X., He, G., & Zhang, F. (2017). Sulfonated poly (phenylene sulfide) grafted polysulfone proton exchange membrane with improved stability. *International Journal of Hydrogen Energy*, 42(4), 2360-2369. DOI: 10.1016/j.ijhydene.2016.09.183

[46] Lobato, J., Canizares, P., Rodrigo, M. A., Úbeda, D., & Pinar, F. J. (2011). Enhancement of the fuel cell performance of a high temperature proton exchange membrane fuel cell running with titanium composite polybenzimidazole-based membranes. *Journal of Power Sources,* 196(20), 8265-8271. DOI: 10.1016/j.jpowsour.2011.06.011

[47] Loureiro, F. A. M., Pereira, R. P., & Rocco, A. M. (2013). Polyethyleneimine-based semi-interpenetrating network membranes for fuel cells. *ECS Transactions*, 58(1), 1153.

[48] Lufrano, F., Squadrito, G., Patti, A., & Passalacqua, E. (2000). Sulfonated polysulfone as promising membranes for polymer electrolyte fuel cells. *Journal of Applied Polymer Science*, 77(6), 1250-1256. DOI: 10.1002/1097-4628(20000808)77:6<1250::aid-app9>3.0.co;2-r

[49] Mohammad, A., & Asiri, A. M. (Eds.). (2017). *Organic-Inorganic Composite Polymer Electrolyte Membranes: Preparation, Properties, and Fuel Cell Applications*. Springer. DOI: 10.1007/978-3-319-52739-0

[50] Neburchilov, V., Martin, J., Wang, H., & Zhang, J. (2007). A review of polymer electrolyte membranes for direct methanol fuel cells. *Journal of Power Sources*, 169(2), 221-238. DOI: 10.1016/j.jpowsour.2007.03.044

[51] Ogungbemi, E., Ijaodola, O., Khatib, F. N., Wilberforce, T., El Hassan, Z., Thompson, J., ... & Olabi, A. G. (2019). Fuel cell membranes–Pros and cons. *Energy*, 172, 155-172. DOI: 10.1016/j.energy.2019.01.034

[52] Ogungbemi, E., Wilberforce, T., Ijaodola, O., Thompson, J., & Olabi, A. G. (2021). Review of operating condition, design parameters and material properties for proton exchange membrane fuel cells. *International Journal of Energy Research*, 45(2), 1227-1245. DOI: 10.1002/er.5810

[53] Park, C. H., Lee, C. H., Guiver, M. D., & Lee, Y. M. (2011). Sulfonated hydrocarbon membranes for medium-temperature and low-humidity proton exchange membrane fuel cells (PEMFCs). *Progress in Polymer Science*, 36(11), 1443-1498. DOI: 10.1016/j.progpolymsci.2011.06.001

[54] Peighambardoust, S. J., Rowshanzamir, S., & Amjadi, M. (2010). Review of the proton exchange membranes for fuel cell applications. *International Journal of Hydrogen Energy*, 35(17), 9349-9384. DOI: 10.1016/j.ijhydene.2010.05.017

[55] Petty-Weeks, S., Zupancic, J. J., & Swedo, J. R. (1988). Proton conducting interpenetrating polymer networks. *Solid State Ionics*, 31(2), 117-125. DOI: 10.1016/0167-2738(88)90295-0

[56] Pu, H., Wang, L., Pan, H., & Wan, D. (2010). Synthesis and characterization of fluorine-containing polybenzimidazole for proton conducting membranes in fuel cells Part A *Polymer Chemistry*. DOI: 10.1002/pola.23979

[57] Ramani, V. I. J. A. Y., Kunz, H. R., & Fenton, J. M. (2004). Investigation of Nafion®/HPA composite membranes for high temperature/low relative humidity PEMFC operation. *Journal of Membrane Science*, 232(1-2), 31-44. DOI: 10.1016/j.memsci.2003.11.016

[58] Rath, R., Kumar, P., Unnikrishnan, L., Mohanty, S., & Nayak, S. K. (2020). Current scenario of poly (2, 5-benzimidazole)(ABPBI) as prospective PEM for application in HT-PEMFC. *Polymer Reviews*, 60(2), 267-317. DOI: 10.1080/15583724.2019.1663211

[59] Rikukawa, M., & Sanui, K. (2000). Proton-conducting polymer electrolyte membranes based on hydrocarbon polymers. *Progress in Polymer Science*, 25(10), 1463-1502. DOI: 10.1016/S0079-6700(00)00032-0

[60] Rosli, R. E., Sulong, A. B., Dauda, W. R. W., Zulkifley, M. A., Husain T., Rosli, M. I., Majlan, E. H., Haque, M. A., (2017). A review of high-temperature proton exchange membrane fuel cell (HT-PEMFC) system, *International Journal of Hydrogen Energy*, 42, 14, 9293-9314. DOI: 10.1016/j.ijhydene.2016.06.211

[61] Staiti, P., Lufrano, F., Arico, A. S., Passalacqua, E., & Antonucci, V. (2001). Sulfonated polybenzimidazole membranes—preparation and physico-chemical characterization. *Journal of Membrane Science*, 188(1), 71-78. DOI: 10.1016/S0376-7388(1)00359-3

[62] Shao, Y., Yin, G., Wang, Z., & Gao, Y. (2007). Proton exchange membrane fuel cell from low temperature to high temperature: material challenges. *Journal of Power Sources*, 167(2), 235-242. DOI: 10.1016/j.jpowsour.2007.02.065

[63] Singh, B., Devi, N., Srivastava, A. K., Singh, R. K., Song, S. J., Krishnan, N. N., ... & Henkensmeier, D. (2018). High temperature polymer electrolyte membrane fuel cells with Polybenzimidazole-Ce_0.

$9Gd_0.1P_2O_7$ and polybenzimidazole-$Ce_0.9Gd_0.1P_2O_7$-graphite oxide composite electrolytes. *Journal of Power Sources*, 401, 149-157. DOI: 10.1016/j.jpowsour.2018.08.076

[64] Singh, R., Oberoi, A. S., & Singh, T. (2022). Factors influencing the performance of PEM fuel cells: A review on performance parameters, water management, and cooling techniques. *International Journal of Energy Research*, 46(4), 3810-3842. DOI: 10.1002/er.7437

[65] Smitha, B., Sridhar, S., & Khan, A. A. (2005). Solid polymer electrolyte membranes for fuel cell applications—a review. *Journal of Membrane Science*, 259(1-2), 10-26. DOI: 10.1016/j.memsci.2005.01.035

[66] Souzy, R., & Ameduri, B. (2005). Functional fluoropolymers for fuel cell membranes. *Progress in Polymer Science*, 30(6), 644-687. DOI: 10.1016/j.progpolymsci.2005.03.004

[67] Souquet-Grumey, J., Perrin, R., Cellier, J., Bigarré, J., & Buvat, P. (2014). Synthesis and fuel cell performance of phosphonated hybrid membranes for PEMFC applications. *Journal of Membrane Science*, 466, 200-210. DOI: 10.1016/j.memsci.2014.04.006

[68] Susai, T., Kaneko, M., Nakato, K., Isono, T., Hamada, A., & Miyake, Y. (2001). Optimization of proton exchange membranes and the humidifying conditions to improve cell performance for polymer electrolyte fuel cells. *International Journal of Hydrogen Energy*, 26(6), 631-637. DOI: 10.1016/S0360-3199(00)00096-3

[69] Tang, Q., Yuan, S., & Cai, H. (2013). High-temperature proton exchange membranes from microporous polyacrylamide caged phosphoric acid. *Journal of Materials Chemistry A*, 1(3), 630-636. DOI: 10.1039/C2TA00116K

[70] Tawil, I. H., Bsebsu, F. M., Hareb, F. O., & Matook, A. M. (2007). Fuel Cells–The Energy Key Of Future. *Fuel*, 1980, 2030.

[71] Tazi, B., & Savadogo, O. (2000). Parameters of PEM fuel-cells based on new membranes fabricated from Nafion®, silicotungstic acid and thiophene. *Electrochimica Acta*, 45(25-26), 4329-4339. DOI: 10.1016/S0013-4686(00)00536-3

[72] Tripathi, B. P., & Shahi, V. K. (2011). Organic–inorganic nanocomposite polymer electrolyte membranes for fuel cell applications. *Progress in Polymer Science*, 36(7), 945-979. DOI: 10.1016/j.progpolymsci.2010.12.005

[73] Wainright, J. S., Wang, J. T., Weng, D., Savinell, R. F., & Litt, M. (1995). Acid-doped polybenzimidazoles: a new polymer electrolyte. *Journal of the Electrochemical Society*, 142(7), L121.

[74] Wakizoe, M., Velev, O. A., & Srinivasan, S. (1995). Analysis of proton exchange membrane fuel cell performance with alternate membranes. *Electrochimica Acta*, 40(3), 335-344. DOI: 10.1016/0013-4686(94)00269-7

[75] Wang, G., Lee, K. H., Lee, W. H., Shin, D. W., Kang, N. R., Cho, D. H., ... & Guiver, M. D. (2014). Durable sulfonated poly (benzothiazole-co-benzimidazole) proton exchange membranes. *Macromolecules*, 47(18), 6355-6364. DOI: 10.1021/ma501409v

[76] Wang, K., Yang, L., Wei, W., Zhang, L., & Chang, G. (2018). Phosphoric acid-doped poly (ether sulfone benzotriazole) for high-temperature proton exchange membrane fuel cell applications. *Journal of Membrane Science*, 549, 23-27. DOI: 10.1016/j.memsci.2017.11.067

[77] Wang, W. F., Hao, R. R., Yang, S. L., Jin, J. H., & Li, G. (2013). Synthesis and Characterization of Sulfonated Polybenzoxazole for High Temperature Proton Exchange Membranes. *In Advanced Materials Research* (Vol. 821, pp. 1261-1265). Trans Tech Publications Ltd. DOI: 10.4028

[78] Wong, C. Y., Wong, W. Y., Ramya, K., Khalid, M., Loh, K. S., Daud, W. R. W., ... & Kadhum, A. A. H. (2019). Additives in proton exchange membranes for low-and high-temperature fuel cell applications: a review. *International Journal of Hydrogen Energy*, 44(12), 6116-6135. DOI: 10.1016/j.ijhydene.2019.01.084

[79] Xu, T. (2005). Ion exchange membranes: state of their development and perspective. *Journal of Membrane Science,* 263(1-2), 1-29. DOI: 10.1016/j.memsci.2005.05.002

[80] Yan, L., & Xie, L. (2012). Molecular dynamics simulations of proton transport in proton exchange membranes based on acid-base complexes. *Molecular Interaction*, 335-360.

[81] Yang, J., Shen, P. K., Varcoe, J., & Wei, Z. (2009). Nafion/polyaniline composite membranes specifically designed to allow proton exchange membrane fuel cells operation at low humidity. *Journal of Power Sources*, 189(2), 1016-1019. DOI: 10.1016/j.jpowsour.2008.12.076

[82] Ye, X., Bai, H., & Ho, W. W. (2006). Synthesis and characterization of new sulfonated polyimides as proton-exchange membranes for fuel cells. *Journal of Membrane Science*, 279(1-2), 570-577. DOI: 10.1016/j.memsci.2005.12.049

[83] Yoshida-Hirahara, M., Takahashi, S., Yoshizawa-Fujita, M., Takeoka, Y., & Rikukawa, M. (2020). Synthesis and investigation of sulfonated poly (p-phenylene)-based ionomers with precisely controlled

ion exchange capacity for use as polymer electrolyte membranes. *RSC Advances*, 10(22), 12810-12822. DOI: 10.1039/D0RA01816C

[84] Zaidi, J., & Matsuura, T. (Eds.). (2008). *Polymer membranes for fuel cells*. Springer Science & Business Media. DOI: 10.1007/978-0-387-73532-0

[85] Zeis, R. (2015). Materials and characterization techniques for high-temperature polymer electrolyte membrane fuel cells. *Beilstein Journal of Nanotechnology*, 6(1), 68-83. DOI: 10.3762/bjnano.6.8

[86] Zhai, Y., Zhang, H., Zhang, Y., & Xing, D. (2007). A novel H_3PO_4/Nafion–PBI composite membrane for enhanced durability of high temperature PEM fuel cells. *Journal of Power Sources*, 169(2), 259-264. DOI: 10.1016/j.jpowsour.2007.03.004

[87] Zhang, J., Xie, Z., Zhang, J., Tang, Y., Song, C., Navessin, T., ... & Holdcroft, S. (2006). High temperature PEM fuel cells. *Journal of power Sources*, 160(2), 872-891. DOI: 10.1016/j.jpowsour.2006.05.034

[88] Zhang, L., Chae, S. R., Hendren, Z., Park, J. S., & Wiesner, M. R. (2012). Recent advances in proton exchange membranes for fuel cell applications. *Chemical Engineering Journal*, 204, 87-97. DOI: 10.1016/j.cej.2012.07.103

[89] Zhang, Y., Zhang, H., Bi, C., & Zhu, X. (2008). An inorganic/organic self-humidifying composite membranes for proton exchange membrane fuel cell application. *Electrochimica Acta*, 53(12), 4096-4103. DOI: 10.1016/j.electacta.2007.12.045

[90] Zuo, Z., Fu, Y., & Manthiram, A. (2012). Novel blend membranes based on acid-base interactions for fuel cells. *Polymers*, 4(4), 1627-1644. DOI: 10.3390/polym4041627

3

Tri-objective Optimization of a Hydrogen-fueled Hybrid Power Generation System

Joy Nondy and T.K. Gogoi

Department of Mechanical Engineering, Tezpur University, India
E-mail: joynondy21.jn@gmail.com; tapan_g@tezu.enet.in

Abstract

In this chapter, a hydrogen-powered hybrid power generation system made up of a proton exchange membrane fuel cell (PEMFC) and a regenerative organic Rankine cycle (RORC) is proposed. The performance of the proposed system is evaluated using energy, exergy, and economic (3E) analyses. A tri-objective optimization is then performed on the hybrid system, considering power output, exergy efficiency, and product cost, using the Pareto envelope-based selection algorithm II. The Technique for Order Preference by Similarity to an Ideal Solution (TOPSIS) is used to select the final optimal solution from the Pareto front. According to the optimized results, the output power of the hybrid system is 1545.3 kW, with an exergy efficiency of 51.02% and a product cost of 197.2 USD/h. Furthermore, it is found that after optimization, the power output and exergy efficiency increased by 6.28% and 13.79%, respectively, and the product cost decreased by 6.25%. The PEMFC is observed to be the most vital component, accounting for 89.55% of total exergy destruction and 70.80% of total capital costs. The scattered distribution plots are finally displayed for each of the decision variables, demonstrating that the conflicting objectives are primarily caused by the current density of the PEMFC.

Keywords: PEM fuel cell, regenerative ORC, tri-objective optimization, PESA-II, TOPSIS

3.1 Introduction

Globally, energy demand is rapidly rising, resulting in increased consumption of fossil fuels, which are already limited by nature. It further results in an increase in pollution, which aggravates the issue of global warming. It inspired the scientific community to seek an alternative fuel-based energy conversion system that emits no harmful emissions. In this regard, fuel cells have gained significant traction in the scientific community due to their numerous advantages over a fossil-fuel-based energy system. The proton exchange membrane fuel cell (PEMFC) is one of the most promising fuel cells because of its zero emission, quiet operation, quick start, and low operating temperature and pressure. However, one major disadvantage of PEMFC is that it generates a lot of heat as a byproduct. Therefore, the PEMFC must be continuously cooled to maintain the operating temperature; otherwise, the membrane will be damaged due to thermal loading. Several researchers have focused on removing the heat from the PEMFC and using it for cogeneration. Hwang et al. (2010) used the rejected heat of PEMFC to produce hot water. In another study, a thermoelectric generator (TEG) was used to produce electricity from the waste heat (Chen et al., 2011). Similarly, in one of the studies, it was shown that the waste heat of PEMFC can also be used to produce chilled water using an absorption chiller (Zhang et al., 2011).

Some of the researchers have also looked into running an organic Rankine cycle (ORC) with the waste heat from a PEMFC. Zhao et al. (2012) integrated an ORC into a PEMFC and evaluated the thermodynamic performance of the overall system using mathematical model equations. They also performed a fluid selection study for the ORC and found that R123 gives the maximum thermal efficiency (42.84%) as compared to R245fa, R245ca, R236fa, and isobutene. Liu et al. (2020) presented an exergy analysis of a PEMFC-ORC hybrid system. They observed that increasing the stack temperature improved the hybrid system's efficiency. They also noticed that the PEMFC stack is accountable for the majority of the system's exergy destruction. Wang et al. (2021) assessed the performance of a hybrid system that included a PEMFC stack and a zeotropic mixture-operated ORC. They compared the performance of the overall systems using various combinations of the zeotropic mixture and found that R245fa/R123 gave the best performance. Lee et al. (2019) proposed two hybrid systems combining PEMFC and ORC, the first of which recovers heat from the PEMFC stack and the second from both the PEMFC stack and the reformer. They reported that the output power increased by 100 and 140 kW for the first and second systems, respectively, when compared to the standalone PEMFC. They also concluded that R123 gives the best performance as the working fluid.

A few studies also focused on optimizing PEMFC-based hybrid systems to improve performance. Kwan et al. (2018) performed the multi-objective optimization of the PEMFC-TEG system using the non-dominated sorting generating algorithm-II (NSGA-II). They chose the output power of TEG and thermal efficiency as objectives, which resulted in a tradeoff solution. In another study, Khanmohammadi et al. (2021) presented a hybrid system that includes compressed air energy storage, the Rankine cycle, PEMFC, and TEG. They also performed a tri-objective optimization considering exergy efficiency, net power, and charging time. They used TOPSIS to obtain the final operating condition from the array of conditions generated through optimization and found the exergy efficiency, net power, and charging time to be 34%, 1078.6 kW, and 5.28 hours, respectively.

According to the brief literature review, it can be observed that many researchers have investigated the possibility of integrating the basic configurations of the ORC with the PEMFC to improve the overall efficiency. However, some studies have found that the regenerative ORC (RORC) outperforms the basic ORC (Nondy & Gogoi, 2021b; Safarian & Aramoun, 2015). With this motivation, the possibility of integrating a regenerative ORC and a PEMFC is investigated here using energy, exergy, and economic (3E) analyses. The prime objectives of this chapter are given as follows:

- The mathematical models of the PEMFC and RORC are validated separately using results reported in the literature. The output results of the PEMFC-RORC system are then compared to the results of another hybrid system (PEMFC-ORC) studied in Ref. (Zhao et al., 2012), to justify using RORC instead of ORC.
- The performance of the PEMFC-RORC is evaluated at base case operating conditions using energy, exergy, and economic (3E) analyses.
- A tri-objective optimization of the proposed hybrid system is performed using PESA-II, considering net power output, exergy efficiency, and total product cost rate as the objective functions. Then TOPSIS is applied to choose the final operating condition (optimal solution) from the 3D Pareto front while giving equal priority to all three objective functions.
- The benefits of tri-objective optimization are shown by comparing how well the hybrid system works at the reference condition and at the optimal condition.

- The criticality of the decision variable toward the tradeoff solution is also studied using scattered distribution plots.

3.2 System Description

Figure 3.1 shows the plan for the hybrid system made up of PEMFC and RORC. The PEMFC setup includes a compressor, H2 tank, pressure regulator, humidifier, and a PEMFC stack. A compressor is used to increase the air pressure from ambient to fuel cell operating pressure. A pressure regulator lowers the pressure of H2 to a level that is suitable for fuel cell operation. The H2 and air streams are then passed through a humidifier into the fuel cell, where they undergo an electrochemical reaction that generates a current. The unreacted H2 from the anode is recirculated back to the fuel cell stack's inlet for reuse after the reaction is complete. At the stack, the exothermic reaction between H2 and air generates a lot of waste heat, which is then converted into electricity using a RORC setup.

The RORC consists of a turbine, feed heater, condenser, and two pumps. The waste heat from the stack is used to vaporize the organic fluid, which enters the turbine in a vapor-saturated state and expands to generate electricity. At an intermediate pressure, some vapor is withdrawn from the turbine and routed to the feed heater. The remaining vapor is expanded in the turbine to condenser pressure. The vapor then exits the turbine and passes through the condenser, where its state is changed from vapor to saturated liquid. Then the liquid is pumped to the feed heater, where it mixes with the extracted vapor from the turbine. The hot mixture is then again pumped to the fuel cell stack for the next cycle.

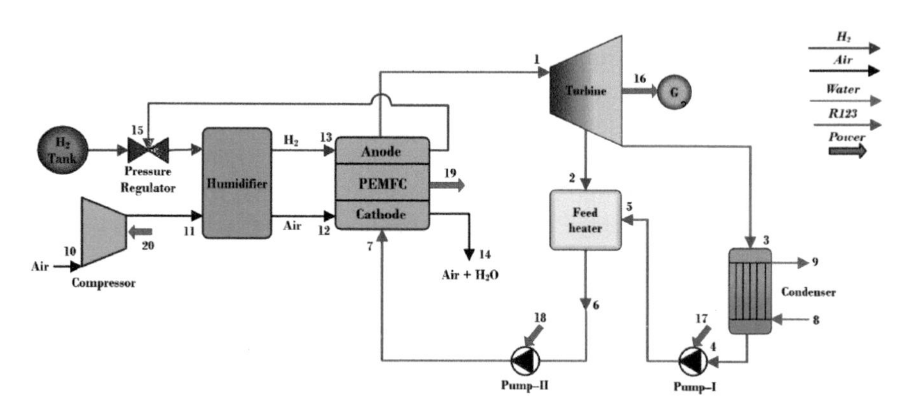

Figure 3.1 Schematic of the hybrid power generation system.

3.2.1 Modeling

This section starts off by discussing the presumptions used to simulate the hybrid system and then moves on to discuss the 3E model equations and multi-objective optimization's framework.

3.2.1.1 Assumptions

The assumptions considered for modeling the hybrid systems are listed below:

- The system is in a steady state (Nondy & Gogoi, 2022).
- Ambient pressure and temperature are 101.3 kPa and 298.15 K, respectively (Nondy & Gogoi, 2020b).
- The composition of ambient air is 79% N_2 and 21% O_2 (Musharavati et al., 2022).
- Organic fluid is used instead of coolant for heat recovery of the fuel cell stack (Zhao et al., 2012).

The input parameters required to model a PEMFC are taken from Ref. (Musharavati et al., 2022) and that of RORC are shown in Table 3.1.

3.2.2 Energy analysis

The governing equations used for modeling the PEMFC and RORC are given in the following sub-sections.

PEM fuel cell: The electrochemical reactions at the fuel cell are given as (Khanmohammadi, Abdi Chaghakaboodi, et al., 2021):

Anode: $H_2 \rightarrow 2H^+ + 2e^-$
Cathode: $2H^+ + 2e^- + \frac{1}{2}O_2 \rightarrow H_2O + \text{Heat}$

Table 3.1 The input parameters of the RORC (Nondy & Gogoi, 2021b).

Parameters	Value	Unit
Turbine isentropic efficiency	80	%
Pump isentropic efficiency	85	%
Turbine inlet temperature	380	K
Condenser temperature	304	K
Vapor extraction pressure	520	kPa
Condensate inlet temperature	298.15	K
Condensate outlet temperature	303.15	K

The net voltage (V_{FC}) produced by the fuel cell due to the reaction is determined using eqn (3.1) (Zhao et al., 2012):

$$V_{\text{FC}} = V_{\text{rev}} - V_{\text{act}} - V_{\text{ohm}} - V_{\text{con}} \tag{3.1}$$

where V_{rev} is the total output voltage termed as reversible cell voltage and V_{act}, V_{ohm}, and V_{con} are the voltage losses due to irreversibilities and termed as activation overvoltage, ohmic overvoltage, and concentration overvoltage, respectively.

The activation overvoltage is estimated using eqn (3.2) (Haghighi & Sharifhassan, 2016).

$$V_{\text{act}} = \left(\frac{\alpha_A + \alpha_C}{\alpha_A \alpha_C}\right) \frac{RT_{\text{FC}}}{nF} \ln\left(\frac{i}{i_0}\right) \tag{3.2}$$

where α_A and α_C are anode's and cathode's transfer coefficient, T_{FC} is the fuel cell's operating temperature, i is the current density, and i_0 is the exchange current density. The exchange current density is calculated by applying the equation given in Ref. (Haghighi & Sharifhassan, 2016).

The ohmic overvoltage is evaluated using eqn (3.3) (Musharavati et al., 2022):

$$V_{\text{ohm}} = i \times t \left[(0.005139\lambda - 0.00326) \exp\left[1286\left(\frac{1}{303} - \frac{1}{T_{\text{FC}}}\right)\right]\right]^{-1} \tag{3.3}$$

where t and λ are the thickness and water content of the membrane, respectively.

The concentration overvoltage is evaluated using eqn (3.4) (Musharavati et al., 2022):

$$V_{\text{con}} = \frac{RT_{\text{FC}}}{nF} \ln\left(\frac{i_L}{i_L - i}\right) \tag{3.4}$$

where i_L is the limiting current density.

Thereafter, the reversible cell voltage is calculated applying eqn (3.5) (Haghighi & Sharifhassan, 2016):

$$\begin{aligned} V_{\text{rev}} = {} & 1.229 - 8.5 \times 10^{-4} (T_{\text{FC}} - 298.15) + 4.3085 \times 10^{-5} \\ & \times T_{\text{FC}} \left[\ln(P_{\text{H}_2}) + \frac{1}{2}\ln(P_{\text{O}_2})\right] \end{aligned} \tag{3.5}$$

where P_{H_2} and P_{O_2} are the partial pressure of H_2 and O_2, which are evaluated by applying correlations reported in Ref. (Musharavati et al., 2022).

Lastly, the electricity developed at the fuel cell stack is given by (Zhao et al., 2012)

$$\dot{W}_{\text{FC}} = N_{\text{cell}} \times V_{\text{FC}} \times i \times A_{\text{cell}} \tag{3.6}$$

where N_{cell} is the number of cells in the stack and A_{cell} is the active surface area of one cell.

The net waste heat ($\dot{Q}_{\text{net}}$) from the stack is evaluated using eqn (3.7) (Musharavati et al., 2022):

$$\dot{Q}_{\text{net}} = \dot{Q}_{\text{ch}} - \dot{W}_{\text{FC}} - \dot{Q}_{s,L} \tag{3.7}$$

where $\dot{Q}_{\text{ch}}$ is the chemical energy and $\dot{Q}_{s,L}$ is the sensible and latent heat. The $\dot{Q}_{\text{ch}}$ and $\dot{Q}_{s,L}$ are evaluated by applying the equations reported in Ref. (Musharavati et al., 2022).

The thermal efficiency of PEMFC is determined using eqn (3.8) (Musharavati et al., 2022).

$$\eta_{\text{PEMFC}} = \frac{\dot{W}_{\text{FC}}}{\dot{n}_{\text{H}_2,\text{cons}} \times \text{HHV}}. \tag{3.8}$$

Regenerative ORC: The RORC is operated with R123 due to lower global warming potential and higher efficiency (Tchanche et al., 2009; Yari & Mahmoudi, 2011). The thermodynamic properties of R123 are estimated using REFPROP 9.0 (Lemmon E, Huber M, n.d.). The state of R123 is assumed as vapor saturated at the inlet of the turbine. The net power ($\dot{W}_{\text{net,ORC}}$) produced by the ORC is evaluated as follows (Zhao et al., 2012):

$$\dot{W}_{\text{net,ORC}} = \dot{W}_T - \dot{W}_{\text{P--I}} - \dot{W}_{\text{P--II}} \tag{3.9}$$

where $\dot{W}_T$ is the power developed at the turbine and $\dot{W}_{\text{P--I}}$ and $\dot{W}_{\text{P--II}}$ are the power delivered to the respective pumps.

The thermal efficiency of the RORC is given by (Shukla et al., 2019)

$$\eta_{\text{RORC}} = \frac{\dot{W}_{\text{net, RORC}}}{\dot{Q}_{\text{net}}}. \tag{3.10}$$

Similarly, the thermal efficiency of the PEMFC-RORC is calculated as follows (Nondy & Gogoi, 2021b):

$$\eta_{\text{sys}} = \frac{\dot{W}_{\text{tot}}}{\dot{n}_{\text{H}_2,\text{cons}} \times \text{HHV}}. \tag{3.11}$$

The energy relation for each component of the proposed system is given in Table 3.2.

Table 3.2 Energy and exergy equations used for modeling PEMFC-RORC.

Components	Energy relations	Exergy relations
Compressor	$\dot{W} = \dot{m}_a\,(h_{11} - h_{10})$	$\dot{E}_{10} + \dot{E}_{20} = \dot{E}_{10} + \dot{E}_D$
PEMFC	$\dot{Q}_{\text{net}} = \dot{m}_r\,(h_1 - h_7)$	$\dot{E}_7 + \dot{E}_{12} + \dot{E}_{13} = \dot{E}_1 + \dot{E}_{14} + \dot{E}_{15} + \dot{E}_{19} + \dot{E}_D$
Turbine	$\dot{W} = \dot{m}_r(h_1 - h_2) + \dot{m}_r\,(1 - x)\,(h_4 - h_5)$	$\dot{E}_1 = \dot{E}_2 + \dot{E}_3 + \dot{E}_{16} + \dot{E}_{17} + \dot{E}_{18} + \dot{E}_D$
	$x = (h_6 - h_5)/(h_2 - h_5)$	–
Condenser	$\dot{m}_r\,(1 - x)\,(h_3 - h_4) = \dot{m}_w\,(h_9 - h_8)$	$\dot{E}_3 + \dot{E}_8 = \dot{E}_4 + \dot{E}_9 + \dot{E}_D$
Pump–I	$\dot{W} = \dot{m}_r\,(1 - x)\,(h_5 - h_4)$	$\dot{E}_4 + \dot{E}_{17} = \dot{E}_5 + \dot{E}_D$
Feed heater	$xh_2 + (1 - x)\,h_5 = h_6$	$\dot{E}_2 + \dot{E}_5 = \dot{E}_6 + \dot{E}_D$
PUMP–II	$\dot{W} = \dot{m}_r\,(h_7 - h_6)$	$\dot{E}_6 + \dot{E}_{18} = \dot{E}_7 + \dot{E}_D$

3.2.3 Exergy analysis

Exergy is the maximum amount of theoretical useful work that can be obtained from a system when it interacts with the dead state. It has four main components, namely physical exergy, chemical exergy, kinetic exergy, and potential exergy. The kinetic and potential exergies are considered zeros since it is assumed that the system under consideration is at rest relative to the environment. Physical exergy is the maximum useful work that can be recovered from a system when it passes from a specific state to a dead state. Chemical exergy is linked to a system's chemical composition deviating from that of the environment. The physical exergy (E^{PH}) is calculated using eqn (3.12) (Souza et al., 2020):

$$\dot{E}^{\text{PH}} = \dot{m}\,[(h - h_0) - T_0(h - s_0)] \tag{3.12}$$

where $\dot{m}$ is the mass flow rate and h and s are the specific enthalpy and entropy.

The chemical exergy O_2 and H_2 are evaluated from the standard molar chemical exergy data reported in Ref. (Bejan et al., 1995). Finally, the exergy balance equation is represented in eqn (3.13) (Musharavati et al., 2022).

$$\dot{E}_F = \dot{E}_P + \dot{E}_L + \dot{E}_D \tag{3.13}$$

where $\dot{E}_F$, $\dot{E}_P$, $\dot{E}_D$, and $\dot{E}_L$ are the rate of fuel exergy, product exergy, exergy destroyed, and exergy loss, respectively.

The exergy balance equation for each component of the proposed system is given in Table 3.2. After determining the exergy rates, the next step is to calculate the exergy efficiency, which indicates the actual performance of each component of a system. The exergy efficiency (ε) can be calculated using eqn (3.14) (Nondy & Gogoi, 2020a):

$$\varepsilon_k = \frac{\dot{E}_{P,k}}{\dot{E}_{F,k}}. \tag{3.14}$$

The exergy efficiencies of PEMFC and RORC are evaluated using eqn (3.15) and (3.16), respectively (Nondy & Gogoi, 2020a).

$$\varepsilon_{\text{PEMFC}} = \frac{\dot{W}_{\text{FC}}}{\dot{E}_{13} + \dot{E}_{12}} \tag{3.15}$$

$$\varepsilon_{\text{sys}} = \frac{\dot{W}_{\text{net, RORC}}}{\dot{E}_1 - \dot{E}_7}. \tag{3.16}$$

3.2.4 Economic analysis

The economic analysis deals with the estimation of expenses such as capital investment costs, fuel costs, and maintenance and operation costs involved in the thermal system. These expenses are calculated to determine the total product cost, which is the expenditures incurred to generate the output products from the thermal system. It is estimated using the cost balance equation as follows:

$$\dot{C}_{P,\text{tot}} = \dot{C}_{F,\text{tot}} + \dot{Z}_{\text{tot}}^{\text{CI}} + \dot{Z}_{\text{tot}}^{\text{OM}} \tag{3.17}$$

where $\dot{C}_{P,\text{tot}}$ is the rate of the total product cost, $\dot{C}_{F,\text{tot}}$ is the rate of the total product cost, $\dot{Z}_{\text{tot}}^{\text{CI}}$ is the capital investment cost rate, and $\dot{Z}_{\text{tot}}^{\text{OM}}$ is the operation and maintenance cost rate.

The cost of getting hydrogen and air is what makes up the total cost of fuel for a PEMFC. The costs of hydrogen and air can be considered 10 $/GJ and 0.011 $/kg, respectively (Kazim, 2005).

The sum of $\dot{Z}_{\text{tot}}^{\text{CI}}$ and $\dot{Z}_{\text{tot}}^{\text{OM}}$ is denoted as $\dot{Z}_{\text{tot}}$, which is referred to as the capital cost rate and for a component k, it can be calculated using eqn (3.18).

$$\dot{Z}_k = \frac{\text{PEC}_k \times \text{CRF} \times \phi}{N} \tag{3.18}$$

where PEC_k stands for purchase equipment cost, ϕ is the maintenance factor (1.06), N is the yearly operation hours (7446 hours) (Khanmohammadi & Azimian, 2015), and CRF stands for capital recovery factor, which is calculated using eqn (3.19).

$$\text{CRF} = \frac{i \times (1+i)^n}{(1+i)^n - 1} \tag{3.19}$$

Table 3.3 The cost equations applied to evaluate PEC_k.

Components	Cost equations	References
Compressor	$\left(\frac{71.10\dot{m}_a}{0.9-\eta_{\text{com}}}\right)\left(\frac{P_{11}}{P_{10}}\right) ln \left(\frac{P_{11}}{P_{10}}\right)$	(Khaljani et al., 2015)
PEMFC	$2500 \times \dot{W}_{\text{PEMFC}}$	(Musharavati et al., 2022)
Turbine	$6000 \left(\dot{W}_{\text{Turbine}}^{0.7}\right)$	(Anvari et al., 2017)
Feed heater	$(527.7/397)^{1.70} \times \acute{C}$ $\log_{10}\acute{C}$ = 4.20-0.204 $\log_{10}\dot{V} + 0.1245\left(\log_{10}\dot{V}\right)^2$	(Anvari et al., 2017)
Condenser	$1773\,(\dot{m}_r)$	(Anvari et al., 2017)
Pumps	$3540 \left(\dot{W}_{\text{Pump}}^{0.71}\right)$	(Anvari et al., 2017)

where i is the rate of interest (12%) and n is the operating period of components (20 years) (Nondy & Gogoi, 2021b).

The PEC_k of each component of the hybrid system is calculated by applying the correlations provided in Table 3.3. The values obtained from the correlations for each component are also updated from their corresponding base year to the reference year using the Marshal and swift cost index (Khaljani et al., 2015).

3.2.5 Optimization

In this study, multi-objective optimization is performed considering total power, overall exergy efficiency, and the product cost rate as the objective functions. The goal is to maximize the total power and the overall exergy efficiency while minimizing the product cost rate. The objective functions are defined as follows.

- Total power:

The total power obtained from the hybrid systems is given by (Zhao et al., 2012)

$$\max : \dot{W}_{\text{tot}} = \dot{W}_{\text{FC}} + \dot{W}_{\text{net,RORC}} \tag{3.20}$$

where $\dot{W}_{\text{tot}}$ is the total power provided by the hybrid system.

- Exergy efficiency:

The exergy efficiency of the overall hybrid system is given by (Anvari et al., 2017)

$$\max : \varepsilon_{\text{sys}} = \frac{\dot{W}_{\text{tot}}}{\dot{E}_{13} + \dot{E}_{12}}. \tag{3.21}$$

Table 3.4 Range of parameters.

Parameters	Unit	Range	References
A_{cell}	cm^2	200–250	(Musharavati et al., 2022)
i	A/cm^2	0.05–1.0	(Haghighi & Sharifhassan, 2016)
t_{mem}	cm	0.016–0.020	(Haghighi & Sharifhassan, 2016)
T_{FC}	K	353.15–373.15	(Zhao et al., 2012)
P_{FC}	kPa	101.32–506.6	(Zhao et al., 2012)
T_1	K	360–400	(Nondy & Gogoi, 2021b)
T_{14}	K	304–308	(Nondy & Gogoi, 2021b)

- Product cost rate:

The product cost rate of the systems is estimated using eqn (3.22) (Nondy & Gogoi, 2020a):

$$\min : \dot{C}_{P,\text{tot}} = \dot{C}_{F,\text{tot}} + \dot{Z}_{\text{tot}} \tag{3.22}$$

where $\dot{Z}_{\text{tot}}$ is the total capital cost rate.

The variables used for the tri-objective optimization are: active surface area (A_{cell}), limiting current density (i), membrane thickness (t_{mem}), fuel cell's operating temperature (T_{FC}) and pressure (P_{FC}), turbine inlet temperature (T_1), and condenser temperature (T_4). The ranges of the decision variables selected based on previously published articles are given in Table 3.4.

The multi-objective optimization is performed using the Pareto envelope-based selection algorithm-II (PESA-II) (Corne et al., 2001). More details on PESA-II can be found in (Nondy & Gogoi, 2021a). In addition, Table 3.5 lists the input parameters required to operate the algorithm. Multi-objective optimization generates a set of non-dominated solutions, also known as Pareto optimal solutions. In the objective space, Pareto optimal solutions are represented by a Pareto front. Though all Pareto optimal solutions are equally good, one solution must be chosen based on the priority of the objective functions. In this study, the best optimal solution is selected using the technique for order preference by similarity to an ideal solution (TOPSIS). TOPSIS is a multi-criteria decision analysis tool that is widely used in problems involving resolving decision dilemmas involving multiple criteria. TOPSIS selects the final solution from an array of options that is nearest to the hypothetical best solution and far away from the hypothetical worst solution. Furthermore, the priority of the objective functions is allotted as weights, with the sum of all weights equal to unity. The algorithm of TOPSIS can be referred to in the previous article of the present authors (Nondy & Gogoi, 2021b).

Table 3.5 The input variables used for running the PESA-II.

Parameters	PESA-II	References
Population size	200	(Nondy & Gogoi, 2021a)
Archive size	200	(Nondy & Gogoi, 2021a)
Crossover probability	0.7	(Nondy & Gogoi, 2021a)
Mutation probability	0.3	(Corne et al., 2001)
Number of iterations	50	(Nondy & Gogoi, 2021a)

3.3 Results and Discussions

3.3.1 Model validation

The hybrid system is modeled using in-house code written in the MAT-LAB environment. The code's accuracy is validated by comparing some key parameters from the current RORC and PEMFC models to published results. Table 3.6 compares some key outputs of PEMFC with outputs reported in Zhao et al. (2012) for identical operating conditions. Similarly, Table 3.7 compares the key outputs of RORC with the results reported in Safarian and Aramoun (2015) for the same input parameters. The results provided by the current models are found to have a marginal deviation from previously published results. Therefore, it confirms the accuracy of the codes developed for simulating the proposed hybrid system.

Table 3.8 depicts the comparative results of the proposed hybrid system (PEMFC-RORC) with another hybrid system (PEMFC-ORC) studied in Ref. (Zhao et al., 2012) under identical operating conditions. It can be observed that the RORC configuration produces 187.3 kW of net power for the same operating condition and working fluid as the basic ORC configuration produces 140.36 kW of net power. As a result, the thermal efficiency of the PEMFC-RORC system is 6.90% higher than that of the PEMFC-ORC system.

Table 3.6 Model validation for PEMFC using the data reported in Ref. (Zhao et al., 2012).

Parameters	Units	Reference work	Current work	Deviation (%)
Total output power	kW	1006.7	1008.7	0.19
Net output voltage	V	0.653	0.666	1.99
Thermal efficiency	%	37.59	37.66	0.18
H_2 consumption rate	mol/s	9.38	9.37	0.10
Air consumption rate	mol/s	22.33	22.31	0.08

Table 3.7 Model validation for RORC using the data reported in Ref. (Safarian & Aramoun, 2015).

Parameters	Units	Reference work	Present work	Deviation (%)
Net output power	kW	55.53	56.55	1.83
Thermal efficiency	%	22.00	22.44	2.00
Mass flow of organic fluid	kg/s	1.99	1.97	1.00
Mass flow of water	kg/s	4.61	4.67	1.30
Exergy destruction	kW	47.30	46.38	1.94

Table 3.8 Comparative results of PEMFC-RORC with PEMFC-ORC.

Parameters	Units	PEMFC-ORC	PEMFC-RORC	Deviation (%)
Turbine power	kW	142.77	192.31	+34.70
Pump power	kW	2.41	5.01	+107.8
Net ORC power	kW	140.36	187.3	+33.44
Thermal efficiency of ORC	%	10.94	14.21	+29.89
Thermal efficiency of the hybrid system	%	42.84	45.80	+6.90

3.3.2 Energy results

The state properties for each nodal point of the hybrid systems at the reference conditions (refer to Table 3.1) are shown in Table 3.9. Additionally, Table 3.10 displays the hybrid system's output parameters discovered by examining how much energy it consumed. It can be seen that the net power obtained from the PEMFC is 1039.1 kW and the power consumed at the compressor is 166.3 kW. Also, PEMFC gives an output voltage of 0.666 V with a thermal efficiency of 38.80%. The molar consumption rates of H_2 and air are 9.37 and 22.31 mol/s, respectively. Next, the net power obtained from the RORC is 187.3 kW with a thermal efficiency of 14.21%. Finally, for the overall hybrid system, the net power obtained is 1226.4 kW with an overall thermal efficiency of 45.8%. It is worth noting that the hybrid system's net power and thermal efficiency are 18.02% and 18.04% higher, respectively, than a standalone PEMFC.

3.3.3 Exergy results

Figure 3.2 shows that PEMFC has the highest exergy destruction of 557 kW. It alone is responsible for 85.7% of the hybrid system's exergy destruction. The presence of irreversibilities caused by electrochemical reactions accounts

Table 3.9 State properties at various state points of the hybrid system.

State	Fluid	T (K)	P (kPa)	$\dot{m}$ (kg/s)	h (kJ/kg)	s (kJ/kgK)	$\dot{Ex}$ (kW)
1	R123	380.00	912.68	8.40	443.30	1.69	352.93
2	R123	362.20	520.00	2.25	435.69	1.69	73.78
3	R123	324.70	112.94	6.16	414.84	1.71	44.52
4	R123	304.00	112.94	6.16	231.13	1.11	1.28
5	R123	304.20	520.00	6.16	231.46	1.11	3.02
6	R123	355.60	520.00	8.40	286.06	1.27	54.79
7	R123	355.89	912.68	8.40	286.41	1.27	57.40
8	Water	298.15	101.33	54.11	104.92	0.37	9.54
9	Water	303.15	101.33	54.11	125.82	0.44	37.73
10	Air	293.15	101.33	1.29	296.15	6.87	0.00
11	Air	419.46	303.96	1.29	424.74	6.92	148.45

Table 3.10 Energy results of the hybrid system at the base case condition.

Parameters	Units	Values
Net output power from PEMFC	kW	1039.1
Power consumed at compressor	kW	166.3
Net output voltage from PEMFC	V	0.666
Thermal efficiency of PEMFC	%	38.80
Molar consumption rate of H_2	mol/s	9.37
Molar consumption rate of air	mol/s	22.31
Net output power from RORC	kW	187.3
Power consumed at pumps	kW	5.01
Thermal efficiency of RORC	%	14.21
Total power	kW	1226.4
Overall thermal efficiency	%	45.8

for large exergy destruction at the PEMFC. The turbine is the next component in the order, with a rate of exergy destruction of 37.27 kW and a share of overall exergy destruction of 5.7%. The feed heater, compressor, and condenser are next, with exergy destruction rates of 22, 17.83, and 15.05 kW, respectively. In comparison to renaming components of the hybrid system, exergy destruction at the pumps is negligible.

Figure 3.3 shows the exergy efficiency of each component of the hybrid system. It is observed that the compressor has the maximum exergy efficiency of 89.27% followed by pump-II (87.64%) and pump-I (85.54%). The turbine and PEMFC have an exergy efficiency of 84.11% and 78.77%, respectively. The condenser has the lowest exergy efficiency since the conversion of fuel exergy into product exergy is poor due to the presence of irreversibilities caused by the heat exchange process. Likewise, due to the presence of

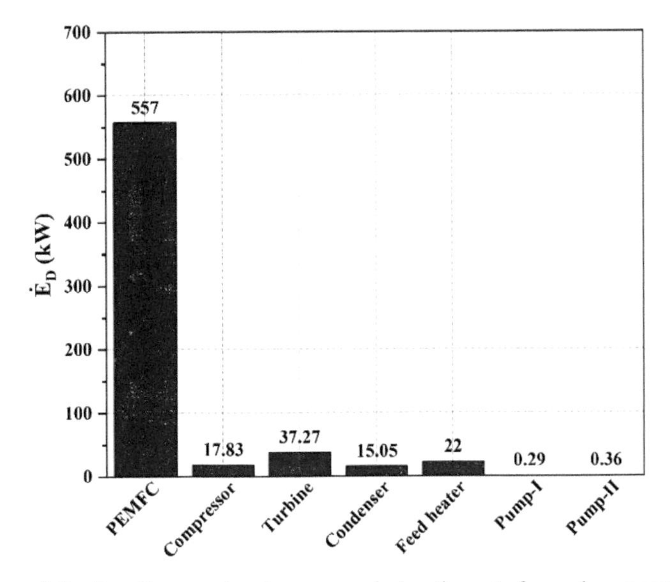

Figure 3.2 Bar diagram showing exergy destruction rate for each component.

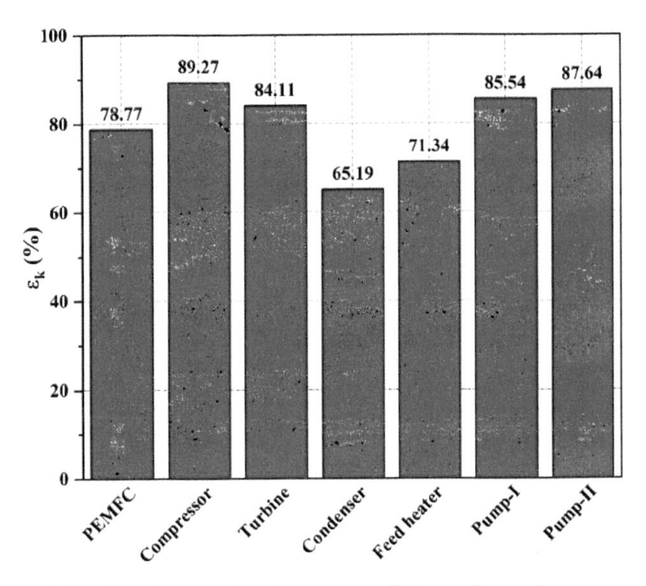

Figure 3.3 Bar diagram showing exergy efficiency for each component.

irreversibilities caused by fluid friction and heat transfer, the exergy efficiency of the feed heater is also relatively low (71.34%).

Table 3.11 Exergy results of the hybrid system at the base case condition.

Parameters	Units	Values
Rate of exergy destruction	kW	649.85
Rate of exergy loss	kW	27.05
Exergy efficiency of fuel cell stack	%	38.56
Exergy efficiency of RORC	%	63.88
Overall exergy efficiency	%	45.52

The rates of exergy destruction and exergy loss of the hybrid system are 649.85 and 27.05 kW, respectively, as shown in Table 3.11. In this study, the exergy leaving the PEMFC along with the stream of air and water (state 14) is considered the exergy loss of the system. Table 3.11 also shows that the exergy efficiencies of PEMFC and RORC are 38.56% and 63.88%, respectively. The overall exergy efficiency of PEMFC-RORC is 45.52%. It implies that the hybrid system could convert 45.5% of the fuel exergy into useful work, particularly electricity in this case.

3.3.4 Economic results

Figure 3.4 shows that PEMFC has the highest capital cost rate of 57.43 $/h. It accounts for nearly 89% of the hybrid system's total capital expenses. The

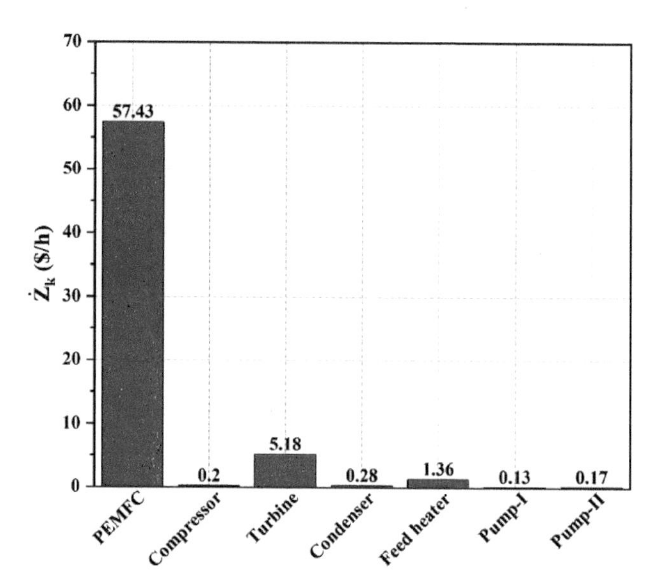

Figure 3.4 Bar diagram showing capital cost rate for each component.

turbine has the next highest capital cost rate of 5.18 $/h followed by a feed heater (1.36 $/h). The capital costs of the remaining components (compressor, condenser, and pumps) are negligible. The total capital cost of hybrid systems is 64.78 $/h, which is determined by adding the individual capital cost rates of each component. The fuel cost rate of the hybrid system is 96.39 $/h. Finally, the product cost rate is evaluated to be 161.18 $/h.

3.3.5 Optimization results

The 3D Pareto front generated by applying PESA-II is depicted in Figure 3.5 (a). The *x*-, *y*-, and *z*-axes represent the power output, exergy efficiency, and total product cost rate, respectively, at the 3D objective space. TOPSIS is used to obtain the final optimal solution by assigning each objective function an equal weight of 1/3. Figure 3.5 (a) also shows the selected optimal solution in the Pareto front. Figure 3.5 (b) shows the top view of the 3D Pareto front, with output power and exergy efficiency represented on the *x*- and *y*-axes, respectively. Similarly, Figures 3.5 (c) and 3.5 (d) show the side and front views of the 3D Pareto front, respectively, with exergy efficiency and total product cost on the *x*-axis and *y*-axis, and output power and total product cost on the *x*-axis and *y*-axis. Figure 3.5 (b) reveals that output power and exergy efficiency are conflicting in nature since, at a higher output power, the exergy efficiency is low. Exergy efficiency and total product cost rate, on the other hand, are compatible in Figure 3.5 (c), as the total product cost rate is low at higher exergy efficiency. Figure 3.5 (d) depicts a conflating relationship between output power and total product cost, with higher output power implying a higher total product cost rate.

Table 3.12 shows the values of the decision variables that correspond to the best optimal solution, and Table 3.13 shows the objective functions calculated at those decision variables. The power output, exergy efficiency, and total product cost rate are 1303.51 kW, 51.80%, and 151.10 $/h, respectively,

Table 3.12 The value decision variables at the best optimal solution.

Decision variables	Unit	Optimal case
A_{cell}	cm^2	248
i	A/cm^2	0.52
t_{mem}	cm	0.0166
T_{FC}	K	353.60
P_{FC}	kPa	104.57
T_1	K	398.91
T_{14}	K	306.55

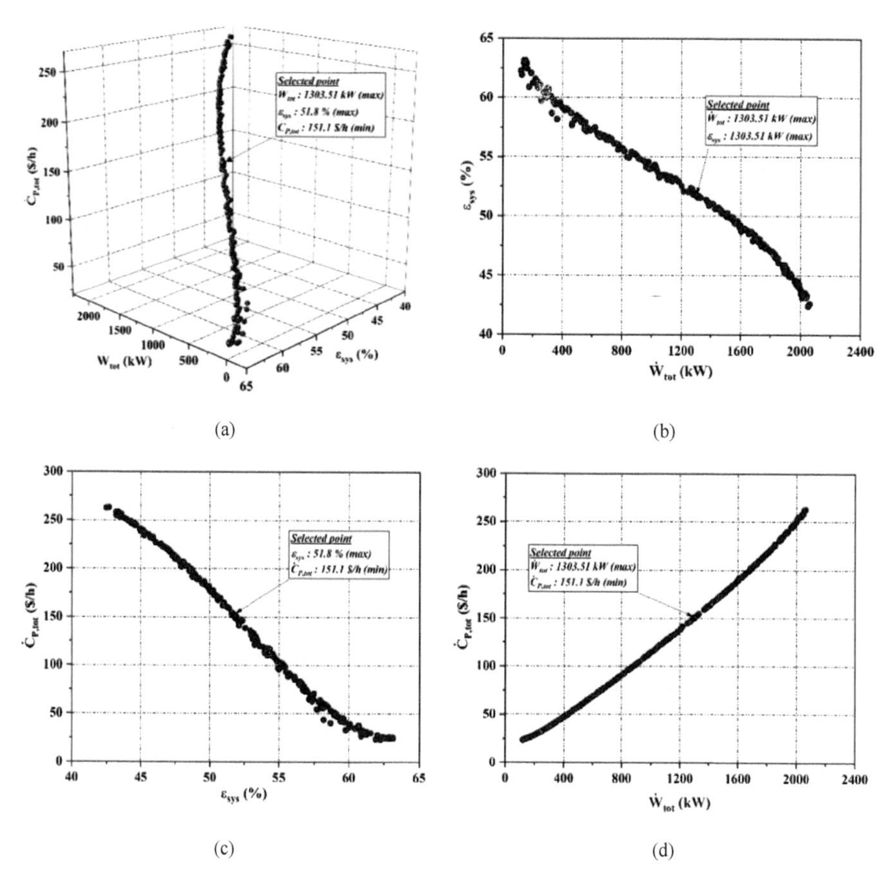

Figure 3.5 Pareto front obtained from the tri-objective optimization of the PEMFC-RORC: (a) 3D view, (b) top view, (c) side view, and (d) front view.

at the optimal condition. The objective functions obtained at the optimal conditions are also compared with that of the base case condition. Table 3.13 shows that after optimizing the hybrid system, the output power and exergy efficiency improve by 6.28% and 13.79%, while the total product cost rate decreases by 6.25%.

Scattered distribution plots are used to show how the decision variables contribute to the conflicting nature of the objective functions. The scattered distribution plots depict the population in the objective space for a particular decision variable. The population distribution reveals the nature of the decision variables concerning the conflicting nature of the objective

Table 3.13 The values objective functions at the best optimal solution.

Objective functions	Unit	Base case	Optimal case	Deviation
$\dot{W}_{tot}$	kW	1226.40	1303.51	+6.28%
ε_{sys}	%	45.52	51.80	+13.79%
$\dot{C}_{P,tot}$	$/h	161.18	151.10	−6.25%

functions. When the population is dispersed across a large objective space, the decision variable is mainly responsible for the tradeoff solution. On the other hand, if the population is densely gathered over a small area, it means that the decision variables complement the objective functions. Figure 3.6 shows the scattered plots for the active cell area. The majority of the population is seen to be dispersed around the upper limits of the objective space. It shows that as the active cell area increases, the three objective functions produce favorable results.

Figure 3.7 shows the scattered plot for the operating temperature of PEMFC. The population is distributed close to the lower bound, implying that the PEMFC's operating temperature has no real impact on the tradeoff solution. It also demonstrates that at a lower value of PEMFC's operating temperature, output power and exergy efficiency are high, and the product

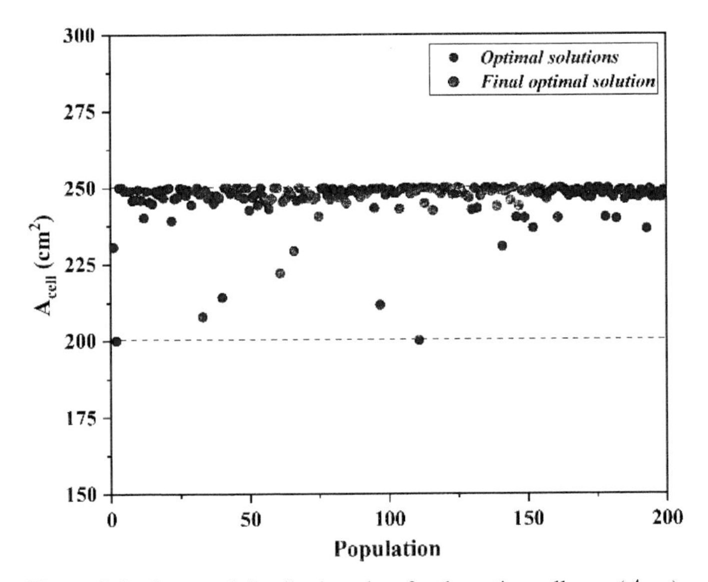

Figure 3.6 Scattered distribution plots for the active cell area (A_{cell}).

cost rate is low. In Figures 3.8 and 3.9, a similar trend can be seen for PEMFC's operating pressure and turbine inlet temperature, respectively.

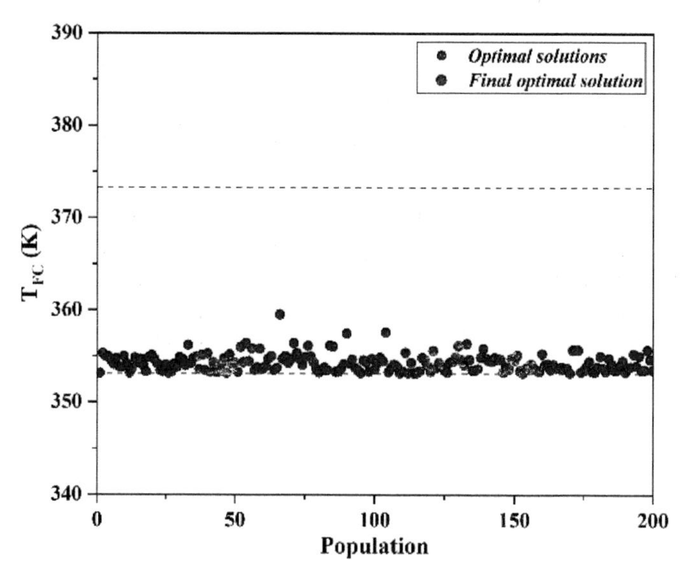

Figure 3.7 Scattered distribution plots for the operating temperature of PEMFC (T_{FC}).

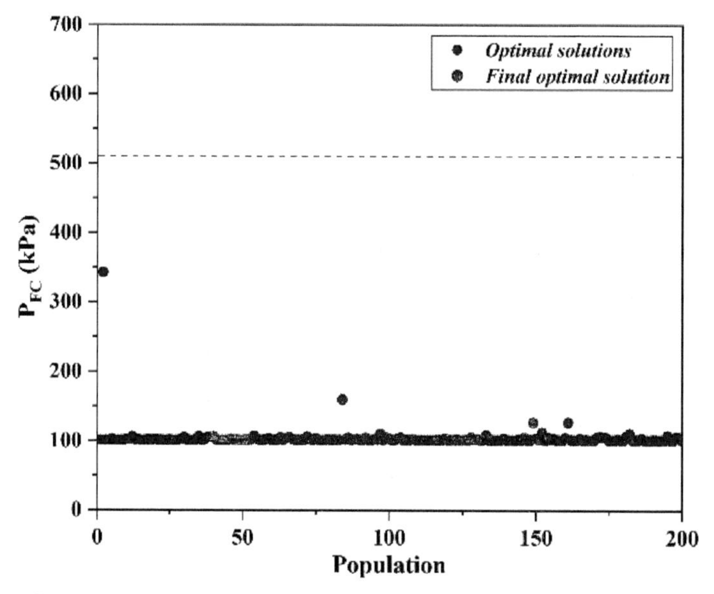

Figure 3.8 Scattered distribution plots for the operating pressure of PEMFC (P_{FC}).

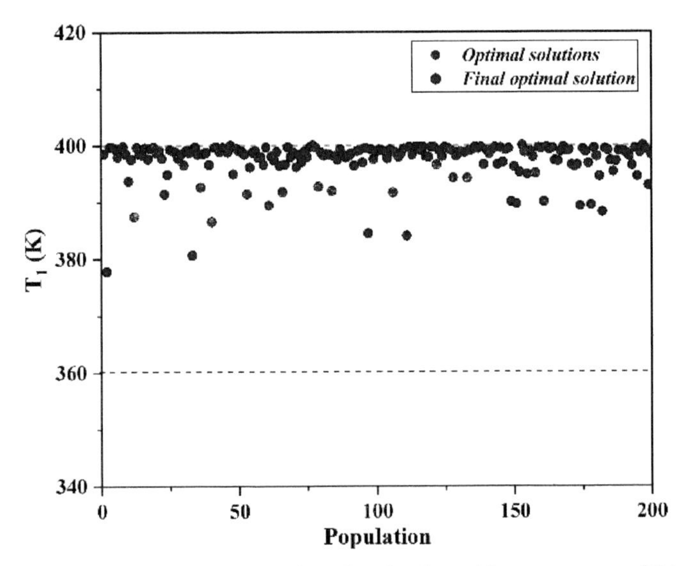

Figure 3.9 Scattered distribution plots for the turbine temperature (T_1).

Figure 3.10 illustrates the scattered plot for the current density. In contrast to the previously discussed decision variables, the population is randomly distributed across the entire objective space in the scattered plot of current density. It implies that current density appreciably influences the tradeoff among objective functions. Any optimal solution that corresponds to current density always results in higher power output, exergy efficiency, and product cost rate. Figure 3.11 shows the scattered plot for the membrane thickness. As can be seen, the population is spared over a region near the lower bound rather than being distributed over the entire objective space. It implies that membrane thickness has a smaller impact on the tradeoff solution than current density.

Lastly, Figure 3.12 shows the scattered plot for the condenser temperature. It can be seen that the population is not distributed evenly across the entire objective space, but rather across the central plain, with a greater number of candidates accumulating in the 306–307.5 K range. It also suggests that, while condenser temperature affects the tradeoff, it has a smaller impact than current density. Therefore, based on the observation of the scattered plots of all the decision variables, it can be concluded that current density is the most critical decision variable that results in the conflicting nature of the objective functions.

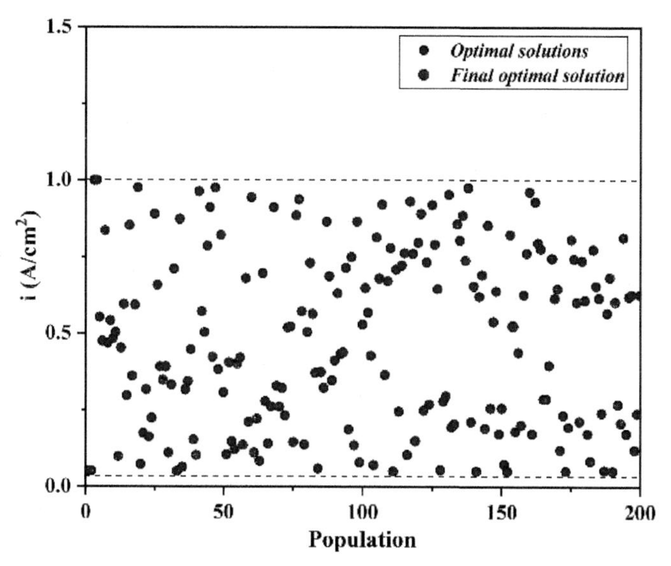

Figure 3.10 Scattered distribution plots for the current density (i).

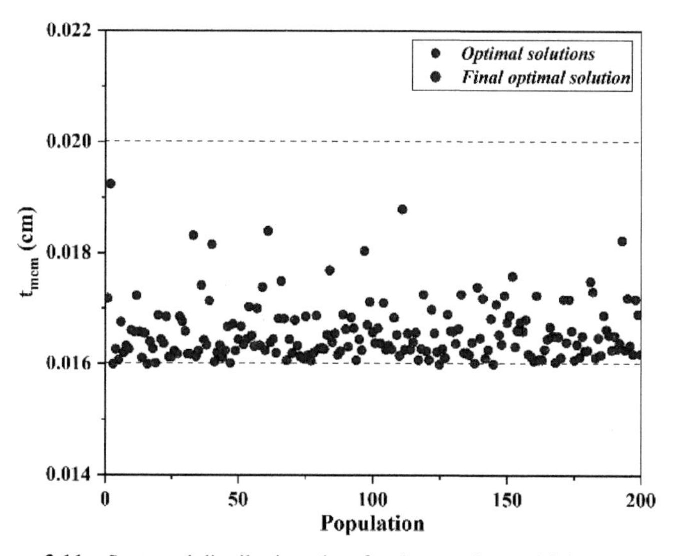

Figure 3.11 Scattered distribution plots for the membrane thickness (t_{mem}).

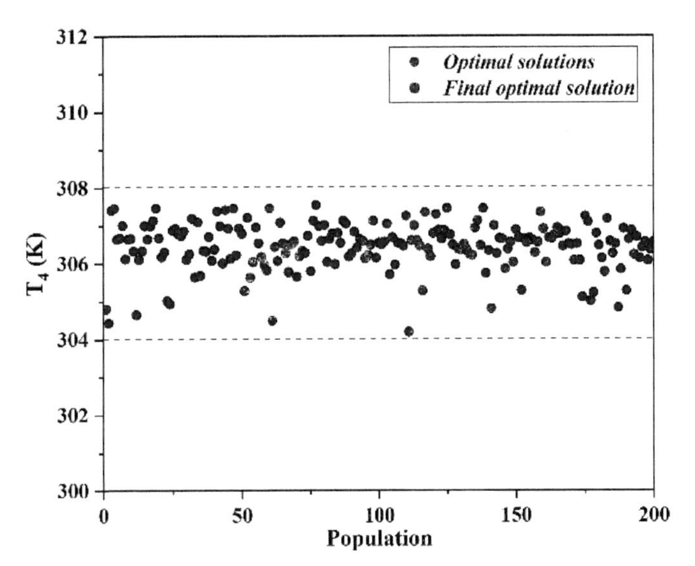

Figure 3.12 Scattered distribution plots for the condenser temperature (T_4).

3.4 Conclusions

This chapter describes a hybrid power system that uses hydrogen and is made up of a PEM fuel cell stack and a regenerative ORC (RORC).The waste heat of the stack is used in the proposed system for operating the RORC. The mathematical model for the proposed system is made in MATLAB, and it is then checked against published results for both PEMFC and RORC. The energy, exergy, and economic (3E) analysis of the hybrid system is then performed under reference working conditions. The thermodynamic results of the hybrid system (PEMFC-RORC) are also compared with another hybrid system (PEMFC-ORC) as reported in Ref. (Zhao et al., 2012). Then PEMFC-RORC system is optimized using PESA-II considering total power, exergy efficiency, and total product cost rate as the objective functions. The final optimal condition of the hybrid system is determined using the TOPSIS decision-maker. The main findings of the study are listed below:

- For the same operating condition and working fluid, the PEMFC-RORC system is found to have 6.9% higher thermal efficiency than the PEMFC-ORC system reported in Ref. (Zhao et al., 2012).
- The proposed hybrid system (PEMFC-RORC) gives a net power of 1226.4 kW with a thermal efficiency of 45.8%. It is worth noting that

the hybrid system's net power and thermal efficiency are 18.02% and 18.04% higher, respectively, than a standalone PEMFC.

- The PEMFC has the highest exergy destruction of 557 kW due to irreversibilities caused by electrochemical reactions.
- PEMFC has the highest capital cost rate of 57.43 $/h accounting for 89% of the overall capital cost. It is followed by turbine and feed heater with the capital cost rate of 5.18 $/h and 1.36 $/h, respectively.
- The output power and exergy efficiency of the hybrid system improved by 6.28% and 13.79%, respectively, while the total product cost rate decreased by 6.25%, according to the performance comparison between the base case and the optimal condition.
- Based on the observation of the scattered plots of all the decision variables, it is observed that current density is the most critical decision variable that results in the conflicting nature of the objective functions.

References

[1] Anvari, S., Taghavifar, H., & Parvishi, A. (2017). Thermo- economical consideration of Regenerative organic Rankine cycle coupling with the absorption chiller systems incorporated in the trigeneration system. Energy Conversion and Management, 148, 317–329. https://doi.org/10.1016/j.enconman.2017.05.077

[2] Bejan, Adrian, George Tsatsaronis, and M. J. M. (1995). Thermal design and optimization. John Wiley & Sons.

[3] Chen, X., Chen, L., Guo, J., & Chen, J. (2011). An available method exploiting the waste heat in a proton exchange membrane fuel cell system. International Journal of Hydrogen Energy, 36(10), 6099–6104. https://doi.org/10.1016/J.IJHYDENE.2011.02.018

[4] Corne, D. W., Jerram, N. R., Knowles, J. D., & Oates, M. J. (2001). PESA-II: Region-based Selection in Evolutionary Multiobjective Optimization. In L. Spector et al. (Ed.), Proc. 3rd Annual Conference on Genetic and Evolutionary Computation (pp. 283–290). https://doi.org/10.5555/2955239.2955289

[5] Haghighi, M., & Sharifhassan, F. (2016). Exergy analysis and optimization of a high temperature proton exchange membrane fuel cell using genetic algorithm. Case Studies in Thermal Engineering, 8, 207–217. https://doi.org/10.1016/j.csite.2016.07.005

[6] Hwang, J. J., Zou, M. L., Chang, W. R., Su, A., Weng, F. B., & Wu, W. (2010). Implementation of a heat recovery unit in a proton exchange membrane fuel cell system. International Journal of Hydrogen Energy, 35(16), 8644–8653. https://doi.org/10.1016/J.IJHYDENE.2010.05.007

[7] J. Souza, R., Dos Santos, C. A. C., Ochoa, A. A. V., Marques, A. S., L. M. Neto, J., & Michima, P. S. A. (2020). Proposal and 3E (energy, exergy, and exergoeconomic) assessment of a cogeneration system using an organic Rankine cycle and an Absorption Refrigeration System in the Northeast Brazil: Thermodynamic investigation of a facility case study. Energy Conversion and Management, 217, 113002. https://doi.org/10.1016/j.enconman.2020.113002

[8] Kazim, A. (2005). Exergoeconomic analysis of a PEM fuel cell at various operating conditions. Energy Conversion and Management, 46(7–8), 1073–1081. https://doi.org/10.1016/j.enconman.2004.06.036

[9] Khaljani, M., Khoshbakhti Saray, R., & Bahlouli, K. (2015). Comprehensive analysis of energy, exergy and exergo-economic of cogeneration of heat and power in a combined gas turbine and organic Rankine cycle. Energy Conversion and Management, 97, 154–165. https://doi.org/10.1016/j.enconman.2015.02.067

[10] Khanmohammadi, S., Abdi Chaghakaboodi, H., & Musharavati, F. (2021). Solar-based Kalina cycle integrated with PEM fuel cell boosted by thermoelectric generator: Development and thermodynamic analysis. International Journal of Green Energy, 1–13. https://doi.org/10.1080/15435075.2021.1881900

[11] Khanmohammadi, S., & Azimian, A. R. (2015). Exergoeconomic Evaluation of a Two-Pressure Level Fired Combined-Cycle Power Plant. Journal of Energy Engineering, 141(3), 04014014. https://doi.org/10.1061/(asce)ey.1943-7897.0000152

[12] Khanmohammadi, S., Rahmani, M., Musharavati, F., Khanmohammadi, S., & Bach, Q. V. (2021). Thermal modeling and triple objective optimization of a new compressed air energy storage system integrated with Rankine cycle, PEM fuel cell, and thermoelectric unit. Sustainable Energy Technologies and Assessments, 43, 100810. https://doi.org/10.1016/J.SETA.2020.100810

[13] Kwan, T. H., Wu, X., & Yao, Q. (2018). Multi-objective genetic optimization of the thermoelectric system for thermal management of proton exchange membrane fuel cells. Applied Energy, 217, 314–327. https://doi.org/10.1016/J.APENERGY.2018.02.097

[14] Lee, J. Y., Lee, J. H., & Kim, T. S. (2019). Thermo-economic analysis of using an organic Rankine cycle for heat recovery from both the cell stack and reformer in a PEMFC for power generation. International Journal of Hydrogen Energy, 44(7), 3876–3890. https://doi.org/10.1016/J.IJHYDENE.2018.12.071

[15] Lemmon E, Huber M, M. M. (n.d.). NIST Standard Reference Database 23, Reference Fluid Thermodynamic and Transport Properties (REFPROP), version 9.0, National Institute of Standards and Technology, R1234yf. fld file dated December 22 (2010). https://www.nist.gov/system/files/documents/srd/REFPROP8_manua3.htm

[16] Liu, G., Qin, Y., Wang, J., Liu, C., Yin, Y., Zhao, J., Yin, Y., Zhang, J., & Nenyi Otoo, O. (2020). Thermodynamic modeling and analysis of a novel PEMFC-ORC combined power system. Energy Conversion and Management, 217, 112998. https://doi.org/10.1016/J.ENCONMAN.2020.112998

[17] Musharavati, F., Khanmohammadi, S., Nondy, J., & Gogoi, T. K. (2022). Proposal of a new low-temperature thermodynamic cycleă: 3E analysis and optimization of a solar pond integrated with fuel cell and thermoelectric generator. Journal of Cleaner Production, 331(August 2021), 129908. https://doi.org/10.1016/j.jclepro.2021.129908

[18] Nondy, J., & Gogoi, T. K. (2020a). A Comparative Study of Metaheuristic Techniques for the Thermoenvironomic Optimization of a Gas Turbine-Based Benchmark Combined Heat and Power System. Journal of Energy Resources Technology, 143(6), 062104. https://doi.org/10.1115/1.4048534

[19] Nondy, J., & Gogoi, T. K. (2020b). Comparative performance analysis of four different combined power and cooling systems integrated with a topping gas turbine plant. Energy Conversion and Management, 223, 113242. https://doi.org/10.1016/j.enconman.2020.113242

[20] Nondy, J., & Gogoi, T. K. (2021a). Performance comparison of multiobjective evolutionary algorithms for exergetic and exergoenvironomic optimization of a benchmark combined heat and power system. Energy, 233, 121135. https://doi.org/10.1016/j.energy.2021.121135

[21] Nondy, J., & Gogoi, T. K. (2021b). Exergoeconomic investigation and multi-objective optimization of different ORC configurations for waste heat recovery: A comparative study. Energy Conversion and Management, 245, 114593. https://doi.org/10.1016/J.ENCONMAN.2021.114593

[22] Nondy, J., & Gogoi, T. K. (2022). Energy and Exergy Analyses of a Gas Turbine and Reheat-Regenerative Steam Turbine Integrated Combined Cycle Power Plant. Lecture Notes in Mechanical Engineering, 233–248. https://doi.org/10.1007/978-981-16-3497-0_18

[23] Safarian, S., & Aramoun, F. (2015). Energy and exergy assessments of modified Organic Rankine Cycles (ORCs). Energy Reports, 1, 1–7. https://doi.org/10.1016/j.egyr.2014.10.003

[24] Shukla, A. K., Sharma, A., Sharma, M., & Nandan, G. (2019). Thermodynamic investigation of solar energy-based triple combined power cycle. Energy Sources, Part A: Recovery, Utilization and Environmental Effects, 41(10), 1161–1179. https://doi.org/10.1080/15567036.2018.1544995

[25] Tchanche, B. F., Papadakis, G., Lambrinos, G., & Frangoudakis, A. (2009). Fluid selection for a low-temperature solar organic Rankine cycle. Applied Thermal Engineering, 29(11–12), 2468–2476. https://doi.org/10.1016/j.applthermaleng.2008.12.025

[26] Wang, C., Li, Q., Wang, C., Zhang, Y., & Zhuge, W. (2021). Thermodynamic analysis of a hydrogen fuel cell waste heat recovery system based on a zeotropic organic Rankine cycle. Energy, 232, 121038. https://doi.org/10.1016/J.ENERGY.2021.121038

[27] Yari, M., & Mahmoudi, S. M. S. (2011). A thermodynamic study of waste heat recovery from GT-MHR using organic Rankine cycles. Heat and Mass Transfer/Waerme- Und Stoffuebertragung, 47(2), 181–196. https://doi.org/10.1007/s00231-010-0698-z

[28] Zhang, X., Chen, X., Lin, B., & Chen, J. (2011). Maximum equivalent efficiency and power output of a PEM fuel cell/refrigeration cycle hybrid system. International Journal of Hydrogen Energy, 36(3), 2190–2196. https://doi.org/10.1016/J.IJHYDENE.2010.11.088

[29] Zhao, P., Wang, J., Gao, L., & Dai, Y. (2012). Parametric analysis of a hybrid power system using organic Rankine cycle to recover waste heat from proton exchange membrane fuel cell. International Journal of Hydrogen Energy, 37(4), 3382–3391. https://doi.org/10.1016/j.ijhydene.2011.11.081

4

Prospects for Hydrogen and Fuel Cells

Ayşenur Öztürk[1] and Ayşe Bayrakçeken Yurtcan[1,2]

[1]Faculty of Engineering, Department of Chemical Engineering, Atatürk University, Turkey
[2]Graduate School of Science, Department of Nanoscience and Nanoengineering, Atatürk University, Turkey
E-mail: aysenur.ozturk@atauni.edu.tr; ayse.bayrakceken@gmail.com

Abstract

Hydrogen-based technologies are getting a lot of attention as clean energy sources that can help meet the growing demand for energy and cut carbon emissions from transportation. The most well-known ways to make hydrogen are through steam reforming (48%), oil reforming (30%), gasification (18%), and electrolysis (4%). Besides hydrogen storage in gaseous (2.70 MJ/L, 30 MPa) and liquid (8.52 MJ/L) forms, the release of hydrogen from metal hydrides (10.8–12.0 MJ/L) is another alternative for hydrogen storage. Hydrogen, which has a high energy density, is used as fuel in fuel cells, which are the top technology products for transportation, stationary use, and portable use. Electric vehicles include battery electric vehicles, fuel cell electric vehicles, fuel cell hybrid electric vehicles, and plug-in fuel cell hybrid electric vehicles. The energy conversion value, which is in the range of 20%–30% in fossil fuel-based vehicles, increases to 50%–60% in vehicles with hydrogen fuel cells. Fuel cell electric vehicles can travel up to 300 miles and have a charging time of fewer than 10 minutes. Today, leading automotive companies are releasing fuel cell-powered vehicle models (Ford Fusion Hydrogen 999, Volvo C30 Electric Range Extender, Chevrolet Sequel, Hyundai ix35 Fuel Cell, Honda Clarity Fuel Cell, Daimler B-Class, and Toyota Mirai). Experts forecast that the number of fuel cell electric vehicles will reach 400 million by 2050. Society will be able to accept hydrogen when the problems of safe storage and transport, lack of infrastructure, high cost, and short-term performance of hydrogen-based products are fixed. Hydrogen's high energy density makes it a good choice for places where fuel tanks are limited in size and weight, like submarines and spacecraft. Notably, hydrogen fuel cells propel the vehicles used in military applications with a tank for long periods, shortening the tank refueling time considerably and reducing the need for maintenance. Hydrogen energy also stands out in aviation with its high gravimetric energy density (33.3 kWh/kg). Fuel cells support the energy grid in buildings and heat the buildings together with the micro-combined heat and pump system. Fuel cells provide 1–30 W of power to portable electronic devices. Hydrogen refueling stations currently offer a capacity of over 1000 kgH_2/day for liquid hydrogen and a range of 100–520 kgH_2/day for gaseous hydrogen. In this chapter, prospects, roadmaps, and product prototypes are discussed regarding hydrogen energy and fuel cells, which will be encountered in all areas of our lives in the future.

Keywords: Hydrogen technologies, fuel cells, fuel cell electric vehicles, energy, prospects, roadmaps

4.1 A Brief Overview of Hydrogen Energy

As the world's population grows and technology improves, there is a great need for new energy sources that can take the place of traditional fossil fuels (FFs). Hydrogen energy (HE) is the leading source among these energy sources, despite some problems that have to be solved. The term hydrogen energy represents the use of the chemical energy of hydrogen itself or hydrogen in compounds to be converted into electrical energy without harming the environment. The specific energy values of the essential energy sources, such as wood, coal, oil, and natural gas (NG), increase with the decreasing carbon amount (Figure 4.1). The specific energy value of hydrogen is higher than all these sources; this is one of the main reasons for hydrogen to be accepted as the future energy star (de Miranda, 2019).

One of the most challenging issues facing humanity is global warming due to greenhouse gases (GHG). Global warming has a devastating effect on the climate and creates changes in the frequency and intensity of abiotic events such as heating, freezing, salinity, and flooding. The major contributor to GHG emissions is carbon dioxide (CO_2) (Zandalinas et al., 2021). Compared to 2000, the Intergovernmental Panel on Climate Change (IPPC) envisages an increase of 25%–90% in total CO_2 emissions and 40%–110% in

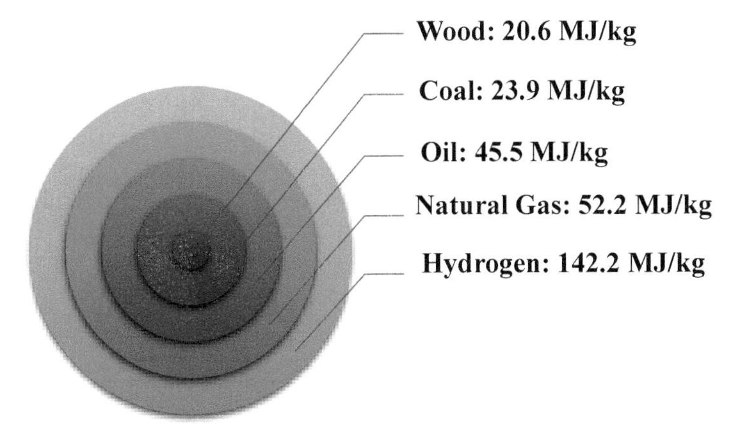

Figure 4.1 Specific energy values of the energy sources (de Miranda, 2019).

CO_2 emissions due to energy consumption by 2030 (Chen et al., 2016). The Kyoto Protocol, signed in 1997 and entered into force in 2005, aims to reduce GHG emissions in line with the personal targets of industrialized countries. The first commitment covered the years between 2008 and 2012 and required countries to reduce their gas emissions by an average of 5% from 1990. The second commitment covered the years between 2013 and 2020 and required countries to reduce their gas emissions by an average of 18% from 1990. As of 2020, global hydrogen production (HP) will have allocated $120.77 billion to the market. With the encouragement of the sanctions determined by the global agreements, the transition to a low-carbon industry will come true with the application of HE into different sectors; thus, the CO_2 emissions will decrease by approximately 6 billion tons. By early 2021, more than 30 governments had shaped the future of energy by defining their hydrogen roadmaps (Pingkuo, 2022). One of the primary objectives of the published roadmaps is to reduce the costs associated with the production, storage, transportation, and safe infrastructure of hydrogen. The World Energy Council Report (2021) publishes the degrees of national hydrogen strategies of the countries (Figure 4.2) (Council, 2021).

The European Climate Foundation (ECF) announced the roadmaps for decreasing GHG emissions by at least 80% compared to 1990 levels and constructing a decarbonized economy from 2010 to 2050. The title of the "Roadmap 2050: A Practical Guide to a Prosperous, Low-Carbon Europe" Declaration consists of some targets as the following (Gandia et al., 2013):

- covering a 5000 km^2 area with solar panels;
- installation and replacement of 100,000 wind turbines;

Figure 4.2 National Hydrogen Strategies (2021) "Used by permission of the World Energy Council" (Council, 2021).

- establishment of an electricity grid system across Europe, securing sustainable electric energy with 190 and 270 GW backup generation systems;
- implementation of a carbon capturing and storage system for the handling of captured CO_2 in the buildings;
- the establishment of more than 100 nuclear power generation facilities by 2040 to meet the 1500 TWh/year of nuclear power generation capacity required for a 40% penetration of renewable energy sources (RES) into the energy sector;
- about 200 million fuel cell electric vehicles (FCEVs) will be used for transportation, and 100 million heat pumps will be put in homes.

Some technical barriers need to be solved for large-scale HE deployment. Reducing high costs is one of the primary goals. Also, hydrogen purification, transmission, and redistribution issues must be handled with safe operating conditions and proper infrastructure (Hu et al., 2020). Despite limitations and poor market conditions, there are many reasons why major automobile manufacturers such as Shell, BMW, Daimler-Chrysler, Ford, Honda, Hyundai, Nissan, General Motors, and Toyota are investing in hydrogen technologies. The materials (FFs, RESs, biomass, water, and nuclear energy) used for HP are diverse, making it theoretically limitless. The use of hydrogen ensures that the energy needs are met independently of the changing energy resources and geographical conditions of the countries and does not pose a risk in terms of national security. Since the energy of hydrogen has been directly converted into chemical energy in fuel cells (FCs), the energy efficiency (EE) is higher than that of internal combustion engines (ICE), limited to the Carnot cycle (Dixon, 2016).

The integrated use of HE with RES (solar, wind, hydraulic, geothermal, tidal, and biomass) increases the utilization of these sources in energy production because they usually work intermittently. The generated power can be converted into different energy forms with this hybridization (Figure 4.3). In addition, it is possible to store the energy produced with hydrogen. RESs are included in the HE chains in four different ways (Maestre et al., 2021):

- **Power to power:** Water electrolysis turns electricity into hydrogen, which can be stored or used in the FC if needed.

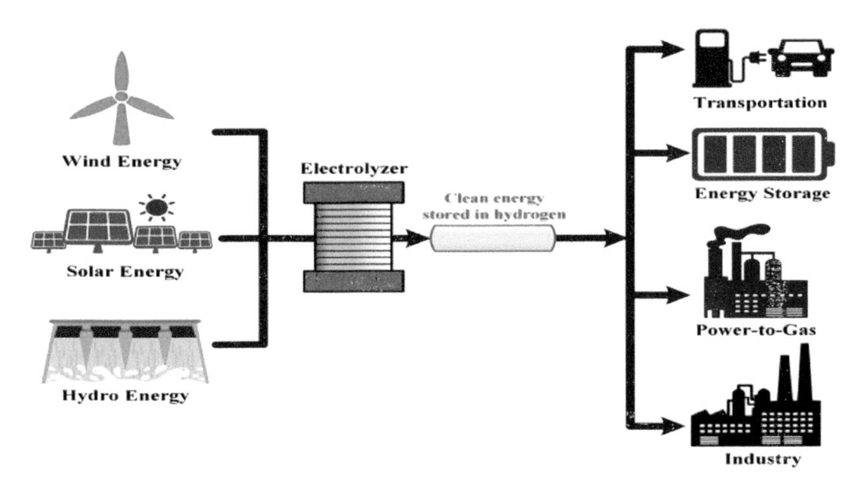

Figure 4.3 Hybridization of water electrolysis with renewable energy sources and utilization patterns of the product hydrogen (reproduced from Ref. (Yodwong et al., 2020) with permission from the MDPI).

- **Power to gas:** Hydrogen converted from electricity via electrolysis can be transported, distributed, and stored directly. It can be also blended into the NG network or converted to synthetic methane.
- **Power to fuel:** Electrolysis turns electricity into hydrogen, which can be used to power FCEVs.
- **Power to feedstock:** Electricity is turned into hydrogen, which is then used as a feedstock to make raw materials, chemical compounds, or synthetic fuels.

4.2 Hydrogen Production

Hydrogen is the most abundant element in the universe; however, it cannot be found in free form and is only found in compounds. For this reason, in systems where hydrogen is an energy carrier, the production method, storage, and compression of hydrogen are issues that must be considered entirely. Even though hydrogen is a clean source of energy, HP methods are not always good for the environment because of the raw materials, energy, and waste they use. The blue, green, and gray colors appear as indicators of the cleanliness of hydrogen, and this situation varies according to the raw material and production method. The green, gray, and blue hydrogen can be obtained from RESs, FFs, and FFs, including CO_2 capturing and sequestering processes, respectively. FFs and RESs are the primary energy sources for HP.

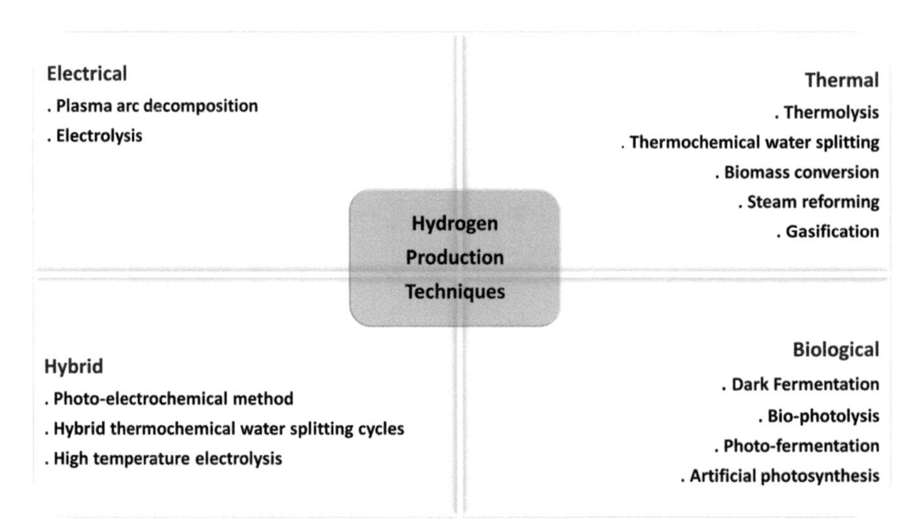

Figure 4.4 Classification of hydrogen production methods (Celik & Yildiz, 2017).

FFs mainly supply the global hydrogen demand, which is 70 million tons per year, although HP from RES is desirable. The leading FF-based production processes are the NG steam reforming process with a share of 48%, while the others are oil reforming (30%) and coal gasification (18%). Water electrolysis and other sources constitute a small share of the remaining 4% (Ji & Wang, 2021). HP methods can be classified into four main groups depending on the input energy source (Figure 4.4) (Celik & Yildiz, 2017).

4.2.1 Thermal techniques

4.2.1.1 Steam reforming

This production process, which started with the reforming of methane in the mid-1990s, has evolved and has been used in HP in different ways: oxidative methanol-reforming, the reforming of sunflower oil over Ni/Al catalysts, *n*-butane reforming, nuclear-heated steam reforming of FFs, the ethanol reforming, the steam reforming of glycerol, the biomass steam gasification, and the NG steam reforming (Barnoon et al., 2021). This method has the highest share in HP techniques since its technology is mature and is suitable for mass HP. In steam reforming, hydrogen is produced from methane gas, which constitutes 98% of NG, by an endothermic process accompanied

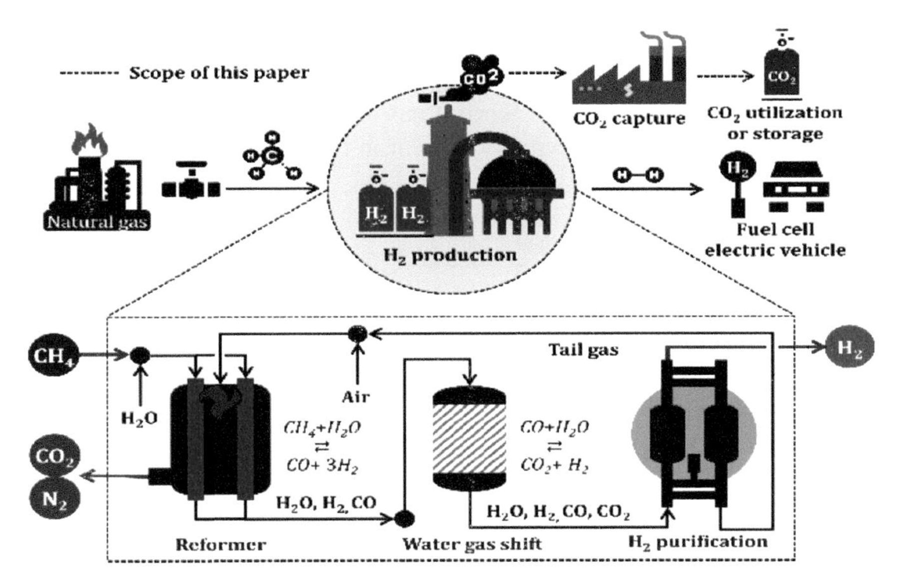

Figure 4.5 Industrial hydrogen production from steam methane reforming (reproduced from Ref. (Lee et al., 2021) with permission from the MDPI).

by Ni-based catalysts. The partial pressures of methane and steam and the reaction temperature affect the hydrogen yield (Ayodele et al., 2021). The reaction is as shown by eqn (4.1):

$$CH_4(g) + H_2O(g) \rightarrow CO(g) + 3H_2(g) \Delta H = 206 \text{ kJ/mole}. \tag{4.1}$$

According to this reaction, syngas ($CO + H_2$) has emerged from liquid or gaseous fuels. The temperature range is 500–900 °C, and the pressure range is between 25 and 30 atm in the reaction environment. Figure 4.5 shows a representative picture of HP from industrial steam reforming (Lee et al., 2021).

4.2.1.2 Gasification

Gasification is the conversion of any carbon-containing raw material into syngas through air, water vapor, or oxygen. The mixed gas produced at the end of the gasification process is purified from various pollutants and by-products to obtain CH_4, CO, and H_2 gases. There are different gasification techniques: fixed bed gasification, moving bed gasification, fluidized bed gasification, entrained flow gasification, and plasma gasification. In the last two gasification techniques, the temperature is between 1200 and 1700 °C, while it is lower than 1200 °C in other techniques (Midilli et al., 2021). Gasification proceeds through a complex mechanism. CO and CO_2 gases release in multiple steps; CO gas can be managed to produce more hydrogen with the water–gas shift (WGS) reaction, but this reaction also proceeds in equilibrium as shown in eqn (4.2). Since it has not been possible to convert CO gas to 100%, this gas is inevitably present in the environment as waste (Celik & Yildiz, 2017).

$$CO + H_2O \xleftrightarrow{\text{high temperature}} CO_2 + H_2. \tag{4.2}$$

4.2.2 Electrical techniques

4.2.2.1 Electrolysis

Electrolysis is one of the preferred environmental methods for HP, and the EE varies between 60% and 80%. However, large-scale HP by this method is limited, as the oxygen evolution reaction (OER) and hydrogen evolution reaction (HER), which take place at the anode and cathode, respectively, occur with noble metals such as platinum (Pt), palladium (Pd), and ruthenium

(Ru) (Anwar et al., 2021). The OER and HER reactions occurring in the electrochemical decomposition of water are as shown in eqn (4.3) and (4.4):

$$\text{Anode}: \ 2H_2O + O_2 \rightarrow 4H^+ + 4e^- (\text{OER}) \tag{4.3}$$

$$\text{Cathode}: \ 4H^+ + 4e^- \rightarrow 2H_2 (\text{HER}). \tag{4.4}$$

In the alkaline electrolysis cell shown in Figure 4.6., water decomposes into oxygen and hydrogen gases by applying an electric current between two electrodes immersed in the potassium hydroxide solution (Rodriguez & Amores, 2020). The hydroxyl (OH^-) ions pass through the membrane to the anode side after the water splits at the cathode electrode. Half-reactions are given by eqn (4.5) and (4.6):

$$\text{Anode}: \ 2OH^- \rightarrow O_2 + H_2O + 2e^- \tag{4.5}$$

$$\text{Cathode}: \ 2H_2O + 2e^- \rightarrow H_2 + 2OH^-. \tag{4.6}$$

Corrosion changes the way the system works and may happen because the electrolyte in the alkaline electrolysis system is corrosive. Today, anion exchange membranes are used instead of diaphragm membranes (Anwar et al., 2021).

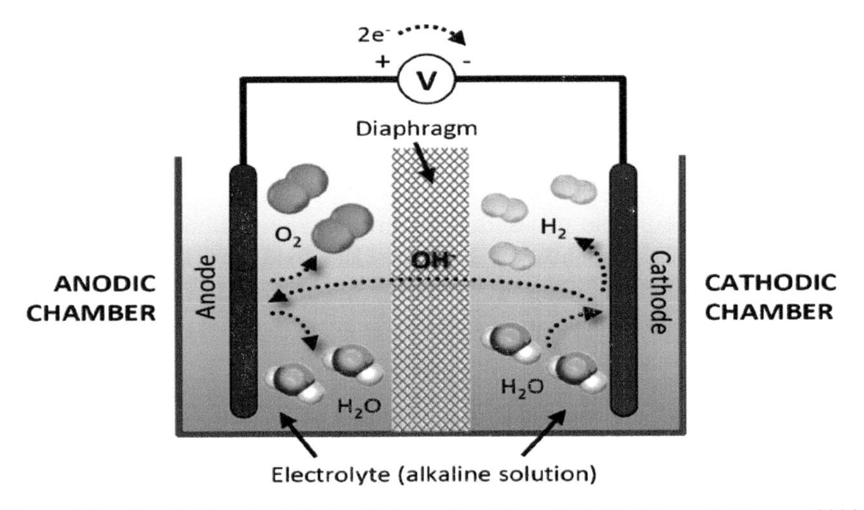

Figure 4.6 An alkaline electrolysis cell (reproduced from Ref. (Rodriguez & Amores, 2020) with permission from the MDPI).

4.2.3 Hybrid techniques

4.2.3.1 Photo-electrochemical method

A photo-electrochemical cell differs from an electrolysis cell in that it contains semiconductor electrodes (at least one of which is covered with a photo-catalyst material). Electrolysis of water is carried out over these semiconductor electrodes in a single cell. A photon hits the photocatalyst material, creates an electron–hole pair, and forms an electrical charge (Celik & Yildiz, 2017). Hydrogen is produced by the reaction shown in eqn (4.7):

$$H_2O(g) \xrightarrow{\text{electricity}} H_2(g) + \frac{1}{2}O_2(g). \tag{4.7}$$

4.2.4 Biological techniques

Fermentative HP takes place through the conversion of biomass (glucose, glycerol, lactate, biowaste, etc.) by microorganisms. Bioconversion can be carried out in two different conditions (Mishra et al., 2019).

4.2.4.1 Dark fermentation

Some of the bacteria that fall into the classes of obligate or facultative anaerobe convert the pure carbon sources or other organic biomass into the hydrogen gas and organic acids. Anaerobes genus Clostridium is the most studied microorganism for HP by this method. HP is limited by the metabolic pathways (Mishra et al., 2019).

4.2.4.2 Photo fermentation

HP occurs in the presence of CO_2 using the light energy from different organic acids and the purple non-sulfur bacterium. This method allows the evaluation of organic acid-rich waste or wastewater for HP. The highest HP efficiency recorded in the literature was 80%. In this method, the wavelengths of light (522, 805, and 850 nm) are the energy sources (Mishra et al., 2019).

4.3 Hydrogen Storage

The market penetration of FCEVs depends on their ability to compete with ICEs in long-range driving. Due to the low density of hydrogen by volume, hydrogen storage (HS) in on-board FCEVs becomes challenging in terms of volume, weight, safety, cost, and reaction kinetics. As of 2025, the volumetric and gravimetric HS capacities have been determined as 5.5 wt.% and 40 g/L

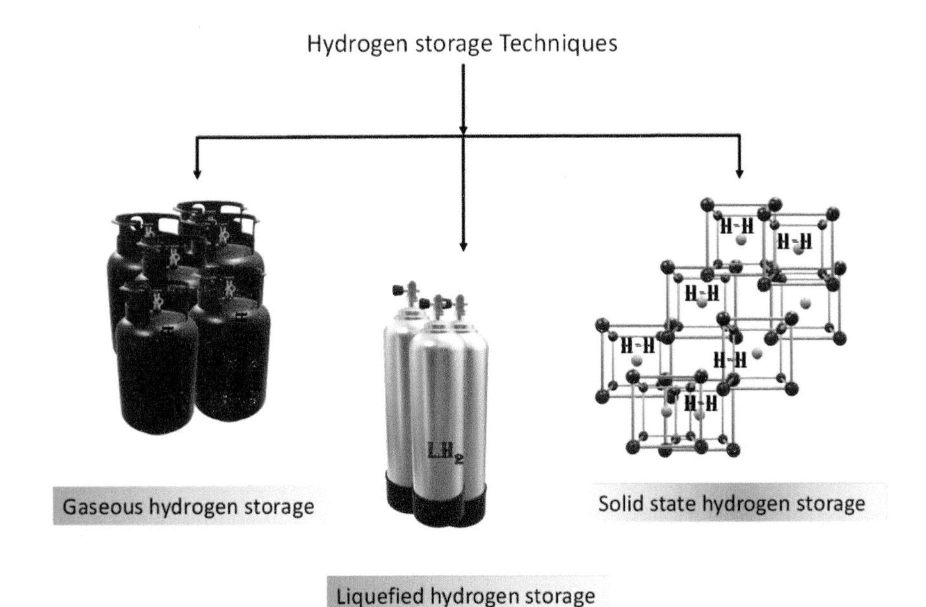

Figure 4.7 Hydrogen storage methods (reproduced from Ref. (Pal, 2021) with permission from the MDPI).

at 233–333 K and 10 MPa pressure by the United States Department of Energy (DOE) (Kudiiarov et al., 2021). There are some methods of HS shown in Figure 4.7 (Pal, 2021).

Gaseous hydrogen storage: Although there is some energy loss in order to compress the hydrogen, it is the approach that FCEV manufacturers prefer (Pollet et al., 2012). The pressurized tanks are in the form of wrapped carbon fiber cylinders with high impact resistance to collisions. 186 L of hydrogen in a pressurized hydrogen tank, which is compressed at 34 MPa and weighs 32.5 kg, provides a traveling distance of 500 km. The problem with these tanks is the volume they occupy in the vehicles (Manoharan et al., 2019).

Liquefied hydrogen storage: Liquid hydrogen has a higher energy density per unit volume, but it needs a very well-insulated tank to stay in liquid form. Liquefaction of hydrogen also causes the loss of 25% of its chemical energy (Pollet et al., 2012). Hydrogen is liquefied at −259.2 °C, which is the cryogenic temperature of hydrogen. Liquid hydrogen can pose a risk of explosion when combined with other gases; so care should be taken when filling hydrogen tanks. The general concept is sweeping the tank with nitrogen gas before filling it with liquid hydrogen. One liter of liquid hydrogen has an electrical energy capacity of 8.52 MJ (Manoharan et al., 2019).

Solid-state hydrogen storage: This method is reliable but cumbersome because the release of hydrogen is so slow that hydrogen refueling takes a long time (Pollet et al., 2012). The metal hydrides (MHs) perform the reversible equilibrium reaction (eqn (4.8)) to give hydrogen:

$$MH_{n,\,\text{solid}} \quad + \quad \text{Heat} \quad \longleftrightarrow \quad M_{\text{solid}} \quad + \quad n/2 \quad H_2 \qquad (4.8)$$

By mixing together different kinds of elements, you can make a wide range of MHs, but there are not many that can be used in car parts. Ideal MH properties for use in cars are a high density, a low dissociation temperature, a low enthalpy for hydrogen release, fast kinetics, recyclability, and a low cost. However, it is very hard to get all of these requirements in one material. Interstitial MHs composed of alloys such as $LaNi_5H_6$ and FeTiH are preferred for light-duty vehicles (LDVs) due to their fast reversible hydrogen charging/discharging rates at ambient conditions. Another class of MHs is complex hydrides such as $LiAlH_4$, $NaAlH_4$, and $NaBH_4$, in which ions form complex bonds with hydrogen. These MHs have higher hydrogen gravimetric densities in the 7%–14% range. Slow hydrogen charging/discharging kinetics are their weakness; so higher temperatures are needed to accelerate the hydrogen dissociation with these MHs (Brooks et al., 2020). The HS method with metal-organic frameworks (MOFs) is an emerging technology. The amount of stored hydrogen varies with temperature, pressure, MOF surface area, and enthalpy of adsorption (Cao et al., 2021). It is possible to store hydrogen on a large scale using liquid organic hydrogen carriers (LOHCs). This method works at room temperature and is recyclable. LOHCs can store hydrogen without binding to any materials or releasing a product, and they can also produce pure hydrogen by dehydrogenation with condensation. Benzene, toluene, naphthalene, N-ethyl carbazole, and dibenzyl toluene are typical LOHC examples (Abdin et al., 2021). Chemical hydrides, especially those in the B-N form (ammonia borane (AB), dimethyl ammonia borane (DAB), and hydrazine borane (HB)), provide a higher energy density than liquid or gaseous HS tanks due to their high hydrogen content. Highly stable and soluble AB (19.6 wt.%) can release hydrogen by hydrolytic dehydrogenation (3 moles of H_2 per mole of AB) in the presence of a suitable catalyst at ambient conditions (Akbayrak & Ozkar, 2018).

4.4 Fuel Cells

FCs are interesting electrochemical devices that directly turn the chemical energy of the fuel into electrical energy. The membrane electrode assembly is the most important part of an FC. It is made up of three parts: an anode, a cathode, and an electrolyte. FCs vary according to the kind of electrolyte, fuel, and operating conditions (Table 4.1). Proton exchange membrane fuel cells (PEMFCs), high-temperature PEMFCs, direct methanol fuel cells (DMFCs), molten carbonate fuel cells (MCFCs), phosphoric acid fuel cells (PAFCs), solid oxide fuel cells (SOFCs), and alkaline fuel cells (AFCs) are the most well-known classes of FCs. PEMFCs have been the pioneers of FC technology because they are FCs that have an acidic character, have a polymeric membrane electrolyte that needs moistening, use Pt as a catalyst, and can be operated at low temperatures (Lucia, 2014).

Although HE penetrated the automotive market with PEMFCs, the high pressure required for HS, the risk of explosion with improper storage conditions, and the flammability of hydrogen led to the development of alternative

Table 4.1 Characteristics of basic fuel cells (reproduced from Ref. (Vaghari, 2013) with permission from the Springer).

Fuel cells	Reactions	Operating temperature (°C)	Electrolyte	Charge carrier
PEMFC & HT-PEMFC	**Anode:** $H_2 \rightarrow 2H^+ + 2e^-$ **Cathode:** $O_2 + 4H^+ + 4e^- \rightarrow 2H_2O$	50–200	Sulfonic acid incorporated into a solid polymer membrane	H^+
MCFC	**Anode:** $H_2 + CO_3{}^{2-} \rightarrow H_2O + CO_2 + 2e^-$ **Cathode:** $O_2 + 2CO_2 + 4e^- \rightarrow 2CO_3{}^{2-}$	630–650	Molten carbonate salt mixture suspended $LiAlO_2$ matrix	$CO_3{}^+$
PAFC	**Anode:** $H_2 \rightarrow 2H^+ + 2e^-$ **Cathode:** $O_2 + 4H^+ + 4e^- \rightarrow 2H_2O$	190–210	Immobilized phosphoric acid in a porous silicon carbide (SiC) matrix	H^+
SOFC	**Anode:** $H_2 + O_2{}^- \rightarrow H_2O + 2e^-$ (for H_2) $CO + O_2{}^- \rightarrow CO_2 + 2e^-$ (for CO) **Cathode:** $O_2 + 4e^- \rightarrow 2O_2{}^-$	700–1000	Ceramic solid oxide zirconia	O_2
AFC	**Anode:** $2H_2 + 4OH^- \rightarrow 4H_2O + 4e^-$ **Cathode:** $O_2 + 2H_2O + 4e^- \rightarrow 4OH^-$	50–200	Molten lithium carbonate	OH^-

hydrogen fuels (Ong et al., 2017). In this sense, direct liquid fuel cells (DLFCs) have formed a large-scale FC class in which liquid fuels containing hydrogen are used as anode electrode feeds. Many liquid fuels, especially alcohols, such as ethanol and methanol, are used in DLFCs. DLFCs impress with advantages like high energy density, fast refueling time, and facile hydrogen transportation and storage. Table 4.2 shows the reactions and energy densities of liquids in DLFCs (Öztürk et al., 2020).

Microfluidic cells (MFCs), which have at least one size in the range of 1–1000 μm, produce bioelectricity sustainably by using microorganisms. The electrons released by the oxidation of the substrate by anaerobe bacteria travel from the anodic chamber to the cathodic chamber and produce electricity on a micro-scale. They are used as low-cost RESs in low-power microelectronics such as sensors, medical implants, and other lab-on-chip devices. The laminar flow of reactants in MFCs removes the necessity of having an ion-exchange membrane in these cell types, thus eliminating problems such as humidification, fuel crossover, and membrane degradation. These cells can be produced economically at room temperature with microfabrication techniques such

Table 4.2 The properties of liquid fuels in direct liquid fuel cells (reproduced from Ref. (Öztürk et al., 2020) with permission from the Elsevier).

Fuel	Abbrv. of DLFCs	Reactions	Energy Density (Wh/L)
Borohydride (B)	DBFC	**Anode:** $BH_4^- + 8OH^- \rightarrow 8BO_2^- + H_2O + 8e^-$ **Cathode:** $2O_2 + 4H_2O + 8e^- \rightarrow 8OH^-$	9946
Dimethyl ether (DE)	DDEFC	**Anode:** $(CH_3)_2O + 3H_2O \rightarrow 2CO_2 + 12H^+ + 12e^-$ **Cathode:** $3O_2 + 12H^+ + 12e^- \rightarrow 6H_2O$	5610
Ethanol (E)	DEFC	**Anode:** $C_2H_5OH + 3H_2O \rightarrow 2CO_2 + 12H^+ + 12e^-$ **Cathode:** $3O_2 + 12H^+ + 12e^- \rightarrow 6H_2O$	6280
Ethylene glycol (EG)	DEGFC	**Anode:** $C_2H_6O_2 + 2H_2O \rightarrow 2CO_2 + 10H^+ + 10e^-$ **Cathode:** $5/2O_2 + 10H^+ + 10e^- \rightarrow 5H_2O$	5800
Formic acid (FA)	DFAFC	**Anode:** $HCOOH \rightarrow CO_2 + 2H^+ + 2e^-$ **Cathode:** $1/2O_2 + 2H^+ + 2e^- \rightarrow H_2O$	1750
Hydrazine (H)	DHFC	**Anode:** $N_2H_4 \rightarrow N_2 + 4H^+ + 4e^-$ **Cathode:** $O_2 + 4H^+ + 4e^- \rightarrow 2H_2O$	5400
Methanol (M)	DMFC	**Anode:** $CH_3OH + H_2O \rightarrow CO_2 + 6H^+ + 6e^-$ **Cathode:** $3/2O_2 + 6H^+ + 6e^- \rightarrow 3H_2O$	4820
1-Propanol (P1)	DP1FC	**Anode:** $CH_3CH_2CH_3OH + 5H_2O \rightarrow 3CO_2 + 18H^+ + 18e^-$ **Cathode:** $9/2O_2 + 18H^+ + 18e^- \rightarrow 9H_2O$	7347
2-Propanol (P2)	DP2FC	**Anode:** $CH_3CH(OH)CH_3 + 5H_2O \rightarrow 3CO_2 + 18H^+ + 18e^-$ **Cathode:** $9/2O_2 + 18H^+ + 18e^- \rightarrow 9H_2O$	7080

as soft lithography, photolithography, and laser etching. Although low performance, long operating times due to low biocatalytic reaction rates, and scale-up problems limit their use, MFCs will take their place in the market with new micro-scale technologies (Parkhey & Sahu, 2021).

4.5 Applications of Fuel Cells

It is possible to make power generation systems (PGSs) for different applications with the desired power output from FC stacks. Developing FC technology has encouraged manufacturers to design low-cost and high-performance PGSs. Although FCs are more prominent in transportation, studies are progressing for developing their markets in other fields. Figure 4.8 shows the three main usage areas of FCs (Vinodh, 2022).

4.5.1 Transport applications

4.5.1.1 Light-duty vehicles

The global stock envisages reaching 2–3 billion units of some LDV models by 2050, which means a serious burden on GHG emissions from transportation. To prevent global warming, electric vehicles (EVs) stand out as the

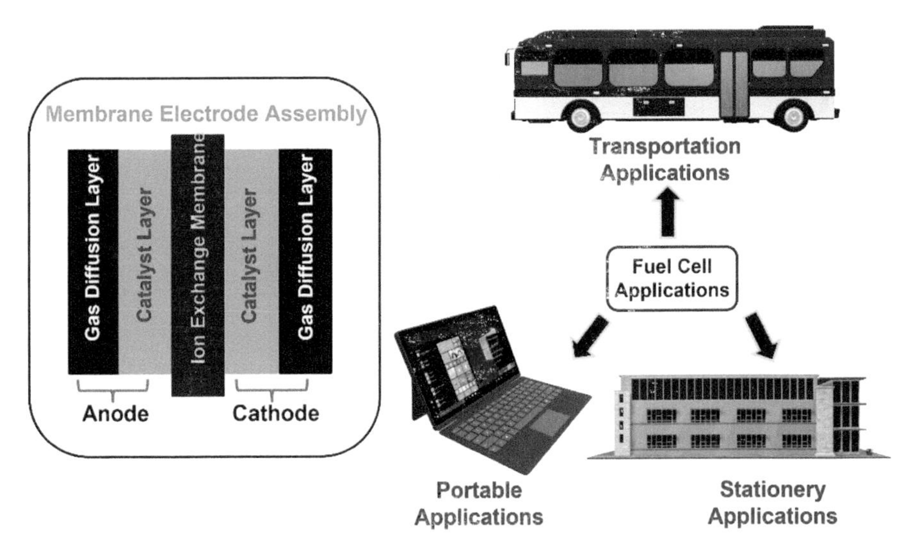

Figure 4.8 Three main branches of fuel cell applications (reproduced from Ref. (Vinodh, 2022) with permission from the MDPI).

future technology in transportation. Automobile manufacturers have turned to different engine designs to reduce storage costs and increase EE in EVs. In technologies called ICE vehicles (ICEVs), EVs, and hybrid electric vehicles (HEVs), as the name suggests, ICEVs work with gasoline-powered engines, while EVs work with a system connected to energy sources such as FCs, batteries, and ultracapacitors. In HEVs, there is a combination of ICEs and electric engines, which depends on the degree of hybridization determined by the electric engine's power to the total engine power ratio. When hybridizing, either a single shaft can be used collectively by the ICE and electric engine or power pathways can be split between the motors. Miles per gallon (MPG) or miles per gallon gasoline-equivalent (MPGe) (33.7 kWh of electrical energy is equal to the energy of one gallon of gasoline) terms define how hybridization improves the fuel economy in HEVs and are used for the comparison of different models in terms of cost (Usai et al., 2021; Das et al., 2017).

EVs whose power propulsion is completely dependent on electrical energy include battery electric vehicles (BEVs), FCEVs, and fuel cell hybrid electric vehicles (FCHEVs) in which FC and battery/ultracapacitor technology are combined (Das et al., 2017). In BEVs, the vehicle-mounted battery is charged intermittently by the grid while the charging system is on-board or off-board. One advantage of these vehicles is that they have less energy loss and low weight due to their simple mechanics. BEVs can be modified with high-traction motors or gearboxes for acceleration, but this reduces the EE of all systems. These vehicles are suitable for urban use because of their low-speed limits, low mileages, and being lightweight. The energy systems are FC stacks fueled with hydrogen in FCEVs. They are called zero-emission vehicles because the by-product is only water and heat. In these systems, hydrogen is either fed from a storage tank or is extracted from a hydrogen-containing fuel using a fuel processor. The distance covered by the FCEVs reaches 300 miles and the refueling times are less than 10 minutes (Manoharan et al., 2019). PEMFCs are favorite in FC stacks because of their high power density and low operating temperature. FCEVs provide constant power but are sensitive to sudden power changes and the system may not respond quickly. FCEVs have been tested in low-speed vehicles such as forklifts, buses, trams, submarines, material handling vehicles, etc.; today, automobile manufacturers are putting them on the market in fast FCEVs and developing engines with different energy management systems (EMSs).

In FCHEVs, FCs are the basic energy source, while batteries or ultracapacitors support the FCs as auxiliary energy storage systems (ESS). The charging and discharging states of batteries or ultracapacitors vary according to the amount of required power. Power converter size, weight, reliability, efficiency, electromagnetic interference (EMI), output voltage and current fluctuation are fundamental technical issues related to FCHEVs' design *(Das et al., 2017)*. One way to reduce dependence on oil is plug-in fuel cell hybrid electric vehicle (PFCHEV) systems, which use the ICE engine in a small part of the power propulsion besides the electrification system. The logic of PFCHEVs is the charging batteries with the grid when the cars are parked overnight and replacing some gasoline with electrical energy *(Thomas, 2009)*. External charging of the ESS increases the range of the car *(Das et al., 2017)*. In the predicted future scenario for LDVs, PFCHEVs will dominate the market due to their extra electrical output for batteries relative to ICEVs and other FCHEVs. Moreover, PFCHEVs that use ethanol instead of oil will further reduce the dependency on FFs *(Thomas, 2009)*. Figure 4.9. shows the simple designs of the powertrains of EVs *(Graber et al., 2022)*.

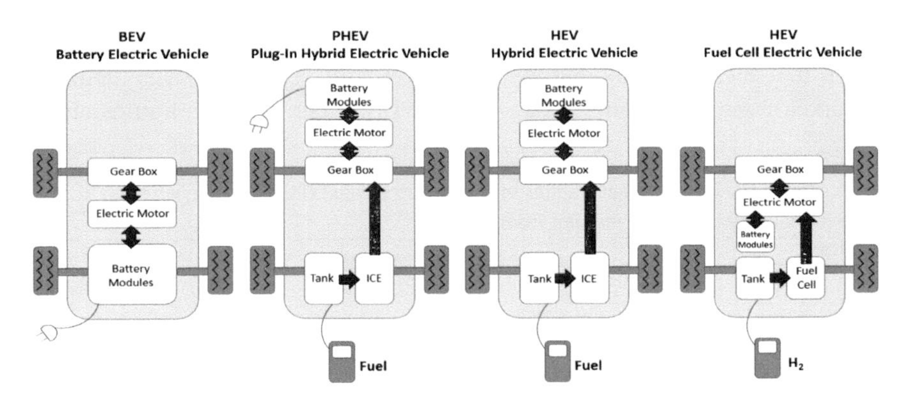

Figure 4.9 Powertrain designs of commercially available electric vehicles (reproduced from Ref. (Graber et al., 2022) with permission from the MDPI).

As of 2019, more than 15,000 FCEVs and FCHEVs have been launched, but among the models, the three that have made a sound are the Toyota Mirai (Japan), Hyundai NEXO (South Korea), and Honda Clarity (Japan), with their respective sales figures (Xun et al., 2021). The Toyota Mirai was introduced to the market in 2014 with large-scale production. This model has become very popular as it is a zero-carbon emission vehicle that can travel 500 km in less than 5 minutes for refueling, and it has become the best-selling FCEV model with a total of 6293 sales (de Almeida & Kruczan, 2021). The breakthroughs necessary for the further commercialization of these vehicles will be possible through technological studies. High EE, long-range driving, and fast refueling are the three targets that the automotive industry wants to achieve in FCEVs (Olabi et al., 2021).

The commercialization of FCEVs around the world requires overcoming some challenges. Although FCEVs are the future of transportation technology, there are still no FCEVs that can compete with conventional ICEVs in terms of economic availability. The principal reason for this is the high cost of FC components and HS equipment (Asif & Schmidt, 2021). Especially, Pt-based catalysts alone account for 40% of the total cost of an FC stack. DOE reduced the percentage of Pt usage in FCs to 0.125 mg/cm^2 for 2020. Other components that increase the cost of the FC after the catalyst are the membrane and bipolar layers. For example, catalyst membranes and bipolar plates account for 80% of the total cost of the FC system for the Toyota Mirai. The DOE has set the current operating cost of FCEVs at $45–60 per kW, but this needs to be made even cheaper (Luo et al., 2021). In Table 4.3, it is seen that the FCEV models released by Toyota, Honda, and Hyundai have market shares in Japan, Europe, and the USA, but the sales figures are

Table 4.3 Fuel cell electric vehicles status in Japan, Europe, and USA (reproduced from Ref. (Asif & Schmidt, 2021) with permission from the MDPI).

Countries	Japan	Europe	USA
Typical products available	Toyota Mirai Honda Clarity (leased only)	Toyota Mirai Honda Clarity (leased only) Hyundai Tucson Hyundai Nexo	Toyota Mirai Honda Clarity (leased only) Hyundai Tucson Hyundai Nexo
Application status	575 and 766 Toyota Mirai were sold in Japan in 2017 and 2018, respectively	130 and 160 Toyota Mirai were sold in Europe in 2017 and 2018, respectively	1700 and 1838 Toyota Mirai were sold in USA in 2017 and 2018, respectively
Fuel cost per kilogram	JPY 1100 (USD 9.85)	EUR 9.50 (USD 11.60)	(USD 16.51)
Infrastructure (number of HRSs)	127(plan to install 160 by fiscal 2021)	177	43(At least 30 more in the stage of planning and construction)
Vehicle cost (Toyota Mirai)	JPY 7.1 million (USD 68.188)	EUR 64,000 (ŸUSD 77.800)	USD 49.500

still low due to high costs. Additionally, despite the increase in the number of hydrogen refueling stations (HRS), this infrastructure opportunity is not homogeneously distributed to all regions of the world (Asif & Schmidt, 2021). Stability and reliability are the two biggest challenges after cost in FCEVs. Especially the lack of long-term stability makes it difficult for end-users to adopt these vehicles. It is troublesome to ensure that every cell and channel in the FC stack operate at the same level. Any fluctuation in flow distribution inside the cells and stacks disrupts the operational balance of the system and leads to low EE. High repair and maintenance costs increase the total cost of FCEVs by up to 60%, resulting in a negative attitude for the end-user (Wang et al., 2018).

4.5.1.2 Heavy-duty vehicles

A heavy-duty vehicle (HDV) has a gross vehicle weight rating (GVWR) greater than 26,000 lbs. These vehicles can be divided into two categories according to their GVWR weights, as shown in Figure 4.10. Vehicles in Class 7 are transit buses, tow trucks, and furniture trucks, while vehicles in Class 8 are semi-tractors, fire trucks, and dump trucks. As with LDVs, HDVs' main objectives are to reduce CO_2 and NOx pollutants released into the air and increase EE in line with low emission strategies. The two implemented strategies for a higher EE with fewer emissions in HDVs are improving current diesel engines or designing new FCs and battery-based engine technologies (Cunanan et al., 2021). FCs provide buses with an operating life of up to 5000 hours, regardless of power generation (Sagaria et al., 2021).

Buses (63%) are of great importance in public transportation in particularly densely populated cities. The average annual journey per capita in countries with populations of 30 million is 40 in the USA and 250 in Japan. The power sources of buses diversify as follows: diesel engines (50%), diesel engines with biodiesel and different additives (22%), electricity (18%), compressed NG, and liquefied petroleum gas (LPG) (10%) (Ajanovic et al., 2021). Hybridization of the power system is common in the general bus prototype (D'Ovidio et al., 2020). In the hybrid power propulsion system, every part of the system contributes to meeting the power demand and solves the problems that may arise in the long run at the same time (Wu et al., 2018). This configuration offers the usage of a high-speed flywheel ESS (FESS) instead of Li-ion batteries to fulfill the higher power request on the road by partitioning the hydrogen FC to avoid transient operation (D'Ovidio et al., 2020). Figure 4.11 shows a prototype of a mid-size transit bus with this hybrid

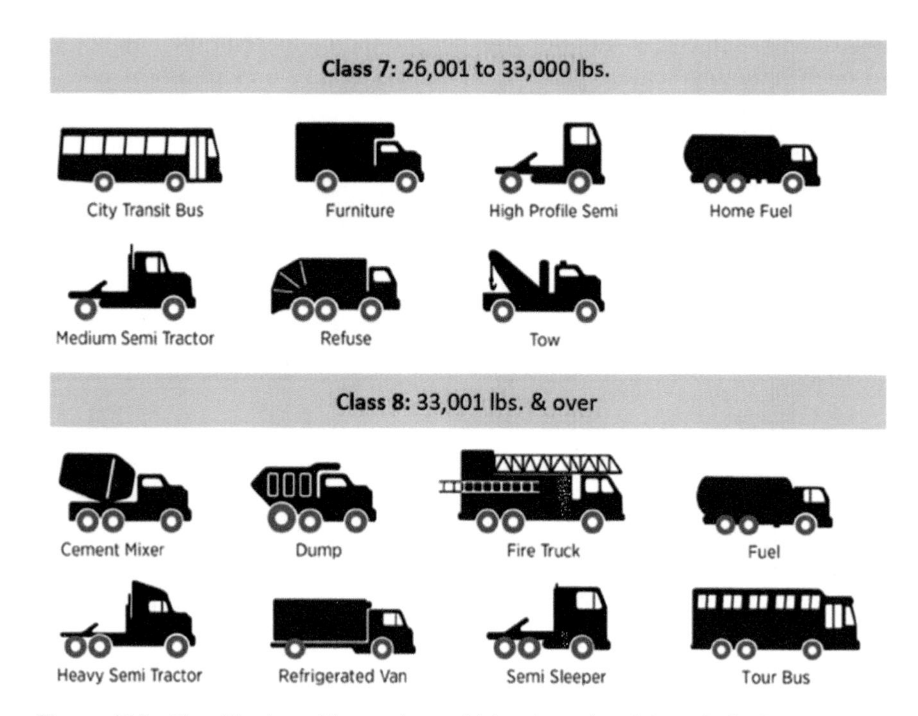

Figure 4.10 Classification of heavy-duty vehicles (reproduced from Ref. (Cunanan et al., 2021) with permission from the MDPI).

propulsion. The energy required to meet the ventilation and air-conditioning requirements in the power systems of passenger transport buses should also be considered.

The buses with diesel and electricity concept started in the 1990s with SCANIA and the German company ZF (Folkesson et al., 2003). Today, various projects are carried out to activate FCs in public vehicles by producing clean energy. Five countries like Germany, the UK, Italy, Latvia, and Denmark aimed to implement 144 FC buses and seven large HRSs within the scope of the JIVE and MEHRLIN projects in 2017 (Europe launches the JIVE and MEHRLIN projects to deliver and refuel 144 fuel cell buses, 2017). In another project, Goldi Mobility collaborated with Hy-Hybrid Energy to produce 12- and 18-m fuel cell buses across the European Union, with Hungary as the central point. These buses will be powered by a 100-kW PEMFC system (Goldi and Horizon plan fuel cell buses for new European project, 2019).

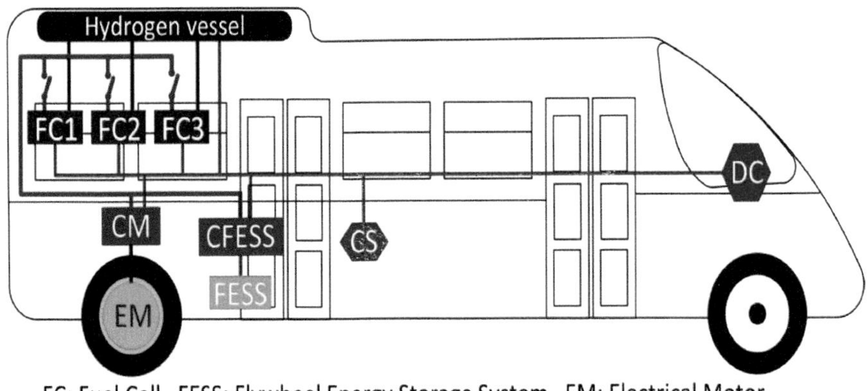

Figure 4.11 Hybrid propulsion of a mid-size transit bus (reproduced from Ref. (D'Ovidio et al., 2020) with permission from the MDPI).

The most successful example of FC applications in the power-to-mobility framework is FC forklifts. There are four different powertrains for forklifts, such as LPG forklifts, diesel forklifts, battery-powered forklifts, and FC-powered forklifts. FC-forklifts provide zero-emission, silent operation, lower refueling time, lower labor cost, and constant power generation (Haghi et al., 2020). HDV trucks need variable power demands depending on structures and operating conditions. Hybrid electric systems and power battery systems are used together in a power configuration (Figure 4.12) to produce the extra power that can meet the heavy load and uphill pulls when these vehicles are loaded. A burning hazard may occur depending on the type of load these vehicles carry in hydrogen supply. Because HDV trucks have limited transit routes and travel times, HRSs for these vehicles may not need to be everywhere (Liu et al., 2020). Hydrogen hybrid powertrain concepts have also been accomplished in railway transportation. The rolling stocks can be categorized into four groups, such as commuter and regional trains, shunting and switching locomotives, mining locomotives, and line haul locomotives; the first group is predominant in terms of electrification with hydrogen fuel cells (Akhoundzadeh et al., 2021). The hybrid system consisting of FCs and ESSs (battery/supercapacitor) is the general configuration in hydrogen trains (Fragiacomo & Piraino, 2019). Fourteen hydrogen-powered

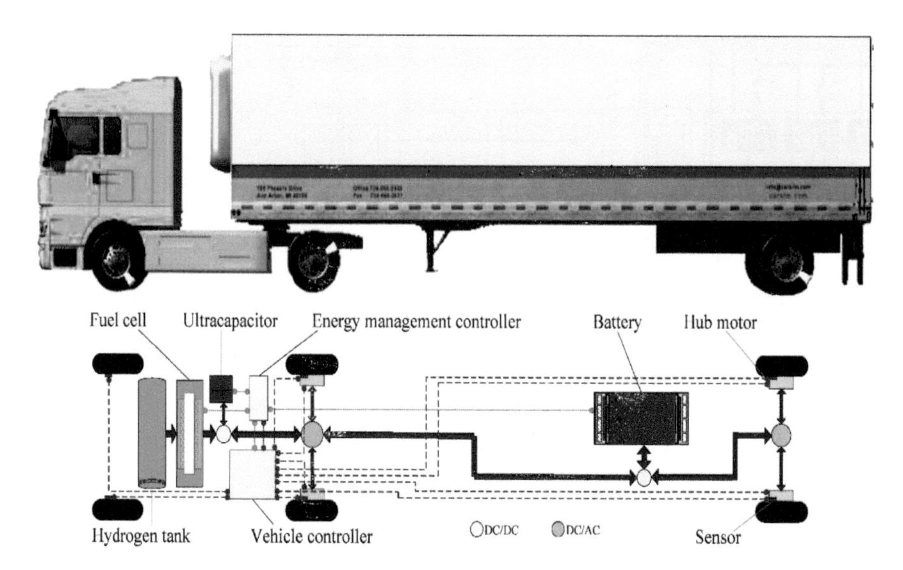

Figure 4.12 Hybrid power system of heavy-duty vehicle truck (reproduced from Ref. (Liu et al., 2020) with permission from the MDPI).

trains, "Coradia iLints," produced by the lead France Alstom company, will operate in Germany in March 2022. Besides Alstom, other manufacturers (Porterbrook, Siemens, Ballard Power Systems, China Railway Rolling Stock Corporation, etc.) have implemented their hydrorail projects in different countries (Palmer, 2022).

Authorities in aviation, like the European Commission's Flightpath 2050, have suggested new energy storage and onboard energy conversion systems to cut fuel use, pollution, and costs in aviation. Hydrogen is an ideal fuel for the aviation industry, with a gravimetric energy density of 33.3 kWh/kg compared to the known fuel AvGas 100LL (12.14 kWh/kg) (Nicolay et al., 2021). The use of hydrogen in aviation is either by using hydrogen as a fuel instead of kerosene in big airplanes or by using FCs instead of jet engines in small airplanes. Airbus and Boeing have experimented with using FCs to power the auxiliary power unit (APU), which functions as the source of electricity and backup system for internal power requirements (cabin pressurization, air conditioning, lighting, etc.). PEMFCs and SOFCs are preferred in aviation; SOFCs also have the advantage of using hydrogen from the kerosene reforming process. Suntan (USA, 1956), Tupolev Tu-155 (Soviet Union, 1988), Cryoplane (Europe, 2000), HyShot (Australia, 2001),

NASA X-43 (USA, 2004), and Phantom Eye (USA, 2013) are exemplary projects for the hydrogen propelled aircraft (Baroutaji et al., 2019).

4.5.1.3 Marine

The emissions from the diesel engines of ships contribute 2.6% CO_2, 15% NOx, and 13% SOx gases to the total GHG emissions, which exceed the share of automobiles. Maritime sanctions have encouraged the use of FCs in this area as well. The elements that make up the configuration of the ship's propulsion are FCs, DC/DC converter, EMS, battery, and propulsion engine. One of the two common types of FCs used in marine applications is PEMFCs, which are preferred in the power system of short-range vehicles such as leisure fishing boats/ferries. The other preferred FC is SOFCs. These FCs are suitable for long-range high-energy applications such as cruise and cargo ships. The Alsterwasser ship of the Zemships project is known as the first FC passenger ship propelled with 48-kW PEMFC and a lead-gel BAT system. This ship achieved a displacement of 72 tons and a top speed of 8 knots in this propulsion system. PEMFCs and SOFCs are generally used in hybrid form with batteries in marine projects that require different power capacities such as submarines, navy ships, passenger ships, megayachts, cruise ships, ferries, sports boats (Ma et al., 2021).

In the early 1980s, two submarine modules working with FCs were designed for the German navy. The first module is 6 kW of AFC technology developed by Siemens with a 100-kW power output. The second module has the propulsion system with PEMFC developed by Siemens in the range of 30–50 kW, to produce an air-independent submarine engine (Strasser, 2010). Electric diesel engines charge batteries while the submarine snorkels; this situation brings with it the risk of detecting the submarine. Air-independent propulsion systems are needed to meet the energy needs of the propulsion and crew accommodation during underwater operations for submarines. This need gave rise to the idea of using FCs in submarines (Psoma & Sattler, 2002). Advantages of FCs over diesel engines:

1. high EE with almost zero emissions;
2. quiet operation;
3. low maintenance cost due to no moving parts;
4. being modular.

Challenges in adapting HE to marine applications include capital and operating costs, HS issues, fuel supply, technology maturity, and safety issues. Increasing the volumetric energy density of hydrogen will prevent difficulties in terms of weight and cost for HS in particular submarines. In addition, increased corrosion due to salt in the air and water, vibration, and water balance should be considered when designing FCs for marine applications (Sürer, 2022).

4.5.1.4 Military

Another potential area for FC technologies is military applications. Modern armed forces are interested in portable devices to increase mobility, capability, and survivability. Being lightweight, compact, and robust during longer operating hours and providing reliable energy flow in harsh environmental conditions are the key points in the design of military portable devices. Devices should withstand different temperature values, such as −40/+70 °C, high humidity, and absorb vibrations and shocks (Cremers, 2009). The defense industry is very suitable for the application of FC technologies, so the investments are continuing in this regard (DMI hydrogen drones for the Korean military, 2021; Intelligent Energy in project to extend flight time in military drones, 2021; NCMS funds a project to integrate FC into Nikola Reckless EV, 2019). The electrification of military vehicles provides reliable operation, accessible power on-board, and high performance. The hybrid ESS can fulfill the need for the high power and high energy capacity of military vehicles due to high-speed operations, cross-country travel, and hilly terrain (Randive et al., 2021).

Drones used for aid, photography, and military purposes need to be able to fly for a long time and do their job well. Wang et al. (2021) used SOFC and PEMFC in the drone power system, and these FCs produced 82.1 and 162.7 W of power, respectively. Different energy generation systems have been tested in unmanned aerial vehicles (UAVs), which have become very popular in recent years. These tools are effective in forest fire management as well as in national security. Solar panels on the wings and FCs are the alternative energy production methods for UAVs. Özbek et al. (2021) designed a mini-UAV called "Hydra" by integrating the PEMFC into this system. Some tests were performed to understand the power requirements of the UAV and to design the proper hybrid power architecture. 160–170 W of the 250 W total power supply to the UAV is propelled by the FC.

4.5.2 Stationary applications

FCs are important in the energy system of buildings. Storing energy in the building, supporting the energy grid as an auxiliary system, or heating the building with a micro-combined heat and pump (CHP) system are the purposes of using FCs. The FCs in residential buildings provide a valuable by-product, such as heat. It also requires a lower power density and ramping rate, has a lower vibration effect, and does not require strict safety measures in the storage tanks (Volkart et al., 2017). The FC technologies for stationary applications are used not only for domestic residences but also for large power generation and industrial facilities. There are examples of energy systems consisting of hybridized FCs and RES in residences. The variable and intermittent nature of RESs brings about the utilization of ESS systems (pumped water, batteries, compressed air, flywheels, supercapacitors, thermal storage, etc.) with these sources. ESSs act as short-term energy storage. HE in buildings can be used standalone or connected to the existing grid (Maestre et al., 2021).

Manufacturers such as Panasonic, Toshiba, Eneos Celltech, Kyocera, Aisin Seiki, and Jxeneos produce FC CHP systems. The utilization of FCs with the CHP configuration increases the performance of the FCs by 60%–90% as a result of heat absorption (Olabi et al., 2020). Another beneficial version of CHP systems is the photovoltaic-thermal (PV/T) module, or solar-thermal collectors, and FC integrated systems. An example of such a system design is shown in Figure 4.13. This configuration aims to increase the total energy and exergy efficiency by including the harvesting of heat from the

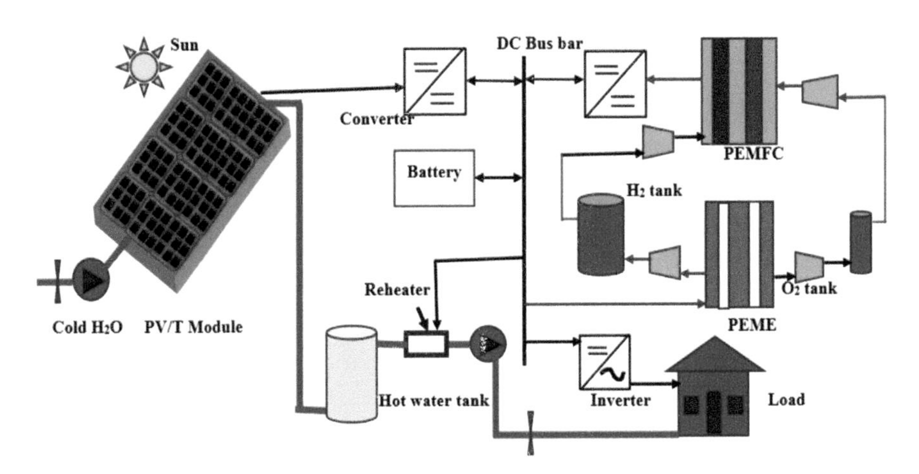

Figure 4.13 Integrated photovoltaic-fuel cell system including photovoltaic-thermal module, battery, electrolyzer, and fuel cell (reproduced from Ref. (Ogbonnaya et al., 2021) with permission from the MDPI).

solar thermal collectors (Ogbonnaya et al., 2021). There are also applications where hydrogen is mixed with NG for cooking or heating purposes. The effect of a mixture of 20% hydrogen mixed with NG on the photovoltaic system has been tested in a real house design in Istanbul/Turkey. The use of this mixture reduced the carbon footprint by 67.5%, while the carbon footprint was reduced by 59.2%, using only NG (Ozcelep et al., 2021).

4.5.3 Mobile applications

FCs can become more popular and last longer if they are tested in real environments and fixed when they do not work well. Performance analyses are made by testing FCs under challenging conditions (start–stop cycle, load–current cycle, heat cycle, and wetting–drying cycle) to understand the degradation mechanisms of FCs and EMSs (Chen et al., 2021). FCs are ideal devices for simple power applications due to their high energy density, durability, simple design, and low cost. Compact and lightweight feed tanks can be used in portable systems to supply fuel continuously to the FC stack. The use of FCs in portable applications is progressing in two main streams. The first is their use as mobile power systems in outdoor activities (camping, climbing, and caravanning) or emergency relief in isolated places with no access to electrical energy. The second-largest area of FCs in mobile applications is consumer electronics such as laptops, tablets, cell phones, radios, camcorders, and ipods, which increase day by day. FCs need to be developed in this area, as the charging of handheld mobile devices and the use of trouble-free power systems are essential in today's increasing communication, digitalization, and entertainment. The power demands of cellular phones, iPhones, personal digital assistants, personal notebooks, tablets, and PlayStations vary in the range of 1–30 W (Wilberforce et al., 2016).

Researchers are working on developing portable power systems that combine HP and FC. Gang and Kwon (2018) developed a portable electric power plant by using a $NaBH_4$-based PEMFC stack. The hydrogen generator in the system produces 5.9 mL H_2/min by catalytic hydrolysis of 20% by mass of $NaBH_4$, which is used as a reactant for the 500-W FC stack. Boran et al. (2019) generated high-purity hydrogen on-demand by hydrolyzing $NaBH_4$. They integrated this system into PEMFC for a portable power device that produces 20 W for charging a notebook for 18 minutes under the optimum hydrolysis conditions. Since soil moisture is of great importance in agriculture, Nguyen et al. (2021) designed soil MFCs to detect soil water holding capacity (SWHC) for 25 days at different moistening rates (40%, 60%, 80%, and 100%). A stable 0.53 V is produced by the soil MFC when the SWHC is above 80%. The authors proposed this system for agriculture as a portable soil moisture sensor.

4.6 Hydrogen Refueling Stations

Most people will be able to use HE because public infrastructure is being built. HRSs act as a bridge between the last link of the HP and the consumption chain (Tian et al., 2022). Depending on how common HE is in the energy market, there will be 4300 large-scale stations and 1500 material-handling stations in the USA by 2030 (Arora, 2022). Apostolou et al. (2021) claimed that a small-scale HRS with a 34.7 €/kgH_2 hydrogen fuel cost (pressurized up to 30 bars) is more favorable than larger stations.

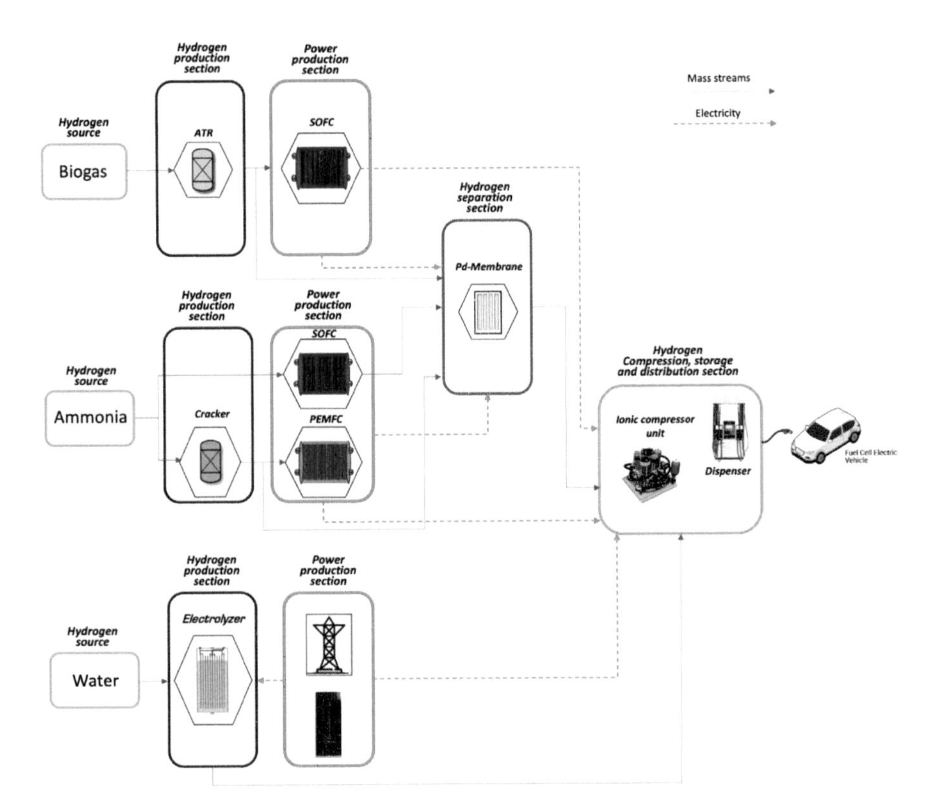

Figure 4.14 Concept design of hydrogen refueling stations (reproduced from Ref. (Perna et al., 2022) with permission from the MDPI).

HRSs are complicated systems with many technical parts, such as compressors, dispensers, high-pressure storage vessels, water chillers, heat exchangers, and so on. The best way to set up and run all of these systems will lower the cost of HE infrastructure and allow FCEVs to compete with ICEVs. Three main factors underlie infrastructure construction: compression, storage, and filling technology. The performance parameters of each piece of equipment greatly affect the overall filling capacity and HS capacity of the HRS (Tian et al., 2022). Figure 4.14 shows the concept design for any HRS. This HRS includes different hydrogen generation sources (ammonia, biogas, and water) and PGS (cracking, autothermal reforming, and electrolysis). The electricity needed for biomass and ammonia-based HP is from PEMFC and SOFC; the energy required for the water electrolysis is supplied from the power generation plant connected to the photovoltaic grid system. A Pd membrane is used to purify the produced hydrogen. Finally, all configurations are connected to the same pressurization, storage, and distribution centers (Perna et al., 2022).

4.7 Conclusions: Conspectus, Prospects, Challenges, and Roadmaps

The following items summarize the issues about hydrogen and fuel cells as conspectus, prospects, challenges, and roadmaps:

Conspectus: (Zhang, 2021), (Inci et al., 2021), (Olabi et al., 2021), and (Apostolou & Xydis, 2019)

Countries that combine RESs with HE will save the future regarding energy demands. In 2020, the global need for hydrogen increased to 324.8 billion. The 90% of HP industries use steam gas reforming. However, the process takes a long time, and the hydrogen purity is around 95%. In the clean electrolysis method, the cost is high and the produced hydrogen is low. In vehicles, PEMFCs have been the most preferred FC. FCEVs have different designs according to the components of their power systems and hybridization types. In LDVs, there are configurations of auxiliary components such as batteries, ultracapacitors, superconducting magnetic energy storage, photovoltaic panels, and flywheels, along with FCs. A wide range of hydrogen buses have been put into operation in Europe. FCs are vital in military applications. RESs that are built with FCs are a smart way to improve EE in homes. On-site HP at HRSs is a desired concept for FCEVs. HRS infrastructures were most abundant in Europe (170), Asia (130), and America (70) at the end of 2018.

Prospects: (Tarkowski, 2022), (Kar, 2022), (Tanc et al., 2019), (Olabi et al., 2021), and (Apostolou & Xydis, 2019)

Hydrogen demand in electricity generation is expected to reach 12 TWh in 2030, 301 TWh in 2040, and 626 TWh in 2050. Investment in electrolysis systems will increase from 3.2 to 8.2 GW by 2050. By 2050, the number of FCEVs will reach 400 million. The International Energy Agency (IEA) estimates that FCEVs will be in the 33,400$ bands by 2050. China aims to bring 1 million FCEVs and 1000 HRSs to the market by 2030. The Global Automotive Executive Survey reports such a distribution in vehicles by 2040: BEVs (26%), FCEVs (25%), ICEs (25%), and hybrids (24%). The cost of produced power per Pt in FCs is expected to be 0.2 g/kW. The H_2 Bus European consortium aims to introduce 1000 new hydrogen-fueled buses, of which 600 are targeted by 2023. The Hydrogen Council predicts the need for 16 billion dollars of investment in developing 15,000 HRS infrastructure to supply the demand of 15 million FCEVs by 2030.

Challenges: (Chau, 2022), (Osmieri & Meyer, 2022), (Tong et al., 2022), (Yue et al., 2021), and (Greene et al., 2020)

The cost of producing hydrogen ($16.26/MPGe) is at least five times greater than producing FFs ($3.09/MPGe). The cost of Pt group metals, which are precious metals used as catalysts in PEMFC stacks, only accounts for 40% of the entire system. The global Pt production amount will meet less than 10% of the future need for FCEVs, considering that 0.25 g Pt/kW is required for FCs. The National Renewable Energy Laboratory found that only 22% of onboard FC stacks tested lasted more than 2000 hours, while the goals for durability before 2024 and 2030 are 6000 and 7000 hours. Systems for drilling, moving, and storing hydrogen must be made of materials that are compatible with hydrogen and the products of its reactions. There is a lack of clear legislation on the use of hydrogen technology in many countries.

Roadmaps: (Hienuki et al., 2019), (Moradi & Groth, 2019), (Maniam et al., 2022), (Peera & Liu, 2022), (Khani, 2022), and (Jiang et al., 2022)

1. The cost of FCs should be reduced by employing non-precious metal catalysts. Generally, non-precious metal catalysts consist of iron (Fe), cobalt (Co), manganese (Mn), nickel (Ni), and copper (Cu) metals together with heteroatom-doped (nitrogen (N), phosphorus (P), sulfur (S), and boron (B)) carbon supports. A good design for catalysts will also increase their utilization rate.

2. The stability of the materials should be increased for the long-term operation of FCs. Bimetallic catalysts formed with Pt are used as ternary structures with a third metal atom to increase the durability of catalysts.
3. Novel membranes and adsorbent materials with high selectivity should be developed for hydrogen purification. Alloy membranes or mixed matrix membranes prepared with different materials are the most recent membrane types for hydrogen purification.
4. Clean HP is possible by reducing the cost of the electrolysis process. The utilization of green hydrogen systems, in which the excess energy of alternative energy systems is used to convert the water to hydrogen, is getting more attention. Seawater electrolysis has recently been suggested as a more economically viable method, although there are some disadvantages in the current situation.
5. Reliable FC-based power systems should be developed that can respond to rapid power ratings in vehicles. The design of hybrid energy storage systems consisting of FC, battery, and ultracapacitor and the development of an optimum real-time predictive energy management strategy will increase instant performance.
6. Hydrogen needs to be stored reliably. Steel pressure vessels with a glass fiber composite overwrap are suitable for utilization in high-pressure conditions. The underground storage of hydrogen is another option.
7. The contribution of each piece of equipment to EE in PGSs of vehicles and buildings should be improved.
8. The design of PGSs with different EMSs should be analyzed beforehand, and the best option should be determined.
9. Sufficient and reliable HRSs and pipelines must be established to refuel FCEVs. The use of tube trailers for hydrogen gas transport has been taken into account by the DOE in recent years, as it requires a simple infrastructure for short distances.
10. Making legal policy and raising public awareness for HE should be realized. The establishment of a reliable and economical hydrogen infrastructure will increase the awareness of hydrogen energy in society.

Acknowledgements

The authors declare that they have no known competing financial interests or personal relationships that could have appeared to influence the work reported in this chapter.

Author Contributions

Ayşenur Öztürk: Investigation, data curation, writing-original draft, conceptualization.

Ayşe Bayrakçeken Yurtcan: Supervision, resources, writing – review and editing.

References

[1] Abdin, Z., Tang, C. G., Liu, Y., & Catchpole, K. (2021). Large-scale stationary hydrogen storage via liquid organic hydrogen carriers. Iscience, 24(9). https://doi.org/ARTN10296610.1016/j.isci.2021.

[2] Ajanovic, A., Glatt, A., & Haas, R. (2021). Prospects and impediments for hydrogen fuel cell buses. Energy, 235. https://doi.org/ARTN121340 10.1016/j.energy.2021.121340

[3] Akbayrak, S., & Ozkar, S. (2018). Ammonia borane as hydrogen storage materials. International Journal of Hydrogen Energy, 43(40), 18592-18606. ttps://doi.org/10.1016/j.ijhydene.2018.02.190

[4] Akhoundzadeh, M. H., Panchal, S., Samadani, E., Raahemifar, K., Fowler, M., & Fraser, R. (2021). Investigation and simulation of electric train utilizing hydrogen fuel cell and lithium-ion battery. Sustainable Energy Technologies and Assessments, 46. https://doi.org/ARTN1012 3410.1016/j.seta.2021.101234

[5] Anwar, S., Khan, F., Zhang, Y. H., & Djire, A. (2021). Recent development in electrocatalysts for hydrogen production through water electrolysis. International Journal of Hydrogen Energy, 46(63), 32284-32317. https://doi.org/10.1016/j.ijhydene.2021.06.191

[6] Apostolou, D., Casero, P., Gil, V., & Xydis, G. (2021). Integration of a light mobility urban scale hydrogen refuelling station for cycling purposes in the transportation market. International Journal of Hydrogen Energy, 46(7), 5756-5762. https://doi.org/10.1016/j.ijhydene.2020.11.0 47

[7] Apostolou, D., & Xydis, G. (2019). A literature review on hydrogen refuelling stations and infrastructure. Current status and future prospects. Renewable & Sustainable Energy Reviews, 113. https://do i.org/ARTN10929210.1016/j.rser.2019.109292

[8] Arora, A., Zantye, M. S., Hasan, M. M. F. (2022). Sustainable hydrogen manufacturing via renewable-integrated intensified process for refueling stations. Applied Energy, 311, Article 118667.

[9] Asif, U., & Schmidt, K. (2021). Fuel Cell Electric Vehicles (FCEV): Policy Advances to Enhance Commercial Success. Sustainability, 13(9). https://doi.org/ARTN514910.3390/su13095149

[10] Ayodele, B. V., Alsaffar, M. A., Mustapa, S. I., Adesina, A., Kanthasamy, R., Witoon, T., & Abdullah, S. (2021). Process intensification of hydrogen production by catalytic steam methane reforming: Performance analysis of multilayer perceptron-artificial neural networks and nonlinear response surface techniques. Process Safety and Environmental Protection, 156, 315-329. https://doi.org/10.1016/j.psep.2021.10.016

[11] Barnoon, P., Toghraie, D., Mehmandoust, B., Fazilati, M. A., & Eftekhari, S. A. (2021). Comprehensive study on hydrogen production via propane steam reforming inside a reactor. Energy Reports, 7, 929-941. https://doi.org/10.1016/j.egyr.2021.02.001

[12] Baroutaji, A., Wilberforce, T., Ramadan, M., & Olabi, A. G. (2019). Comprehensive investigation on hydrogen and fuel cell technology in the aviation and aerospace sectors. Renewable & Sustainable Energy Reviews, 106, 31-40. https://doi.org/10.1016/j.rser.2019.02.022

[13] Boran, A., Erkan, S., & Eroglu, I. (2019). Hydrogen generation from solid state NaBH4 by using FeCl3 catalyst for portable proton exchange membrane fuel cell applications. International Journal of Hydrogen Energy, 44(34), 18915-18926. https://doi.org/10.1016/j.ijhydene.2018.11.033

[14] Brooks, K. P., Sprik, S. J., Tamburello, D. A., & Thornton, M. J. (2020). Design tool for estimating metal hydride storage system characteristics for light-duty hydrogen fuel cell vehicles. International Journal of Hydrogen Energy, 45(46), 24917-24927. https://doi.org/10.1016/j.ijhydene.2020.05.159

[15] Cao, Y., Dhahad, H. A., Zare, S. G., Farouk, N., Anqi, A. E., Issakhov, A., & Raise, A. (2021). Potential application of metal-organic frameworks (MOFs) for hydrogen storage: Simulation by artificial intelligent techniques. International Journal of Hydrogen Energy, 46(73) 36336-36347. https://doi.org/10.1016/j.ijhydene.2021.08.167

[16] Celik, D., & Yildiz, M. (2017). Investigation of hydrogen production methods in accordance with green chemistry principles. International Journal of Hydrogen Energy, 42(36), 23395-23401. https://doi.org/10.1016/j.ijhydene.2017.03.104

[17] Chau, K., Djire, A., Khan, F. (2022). Review and analysis of the hydrogen production technologies from a safety perspective. International Journal of Hydrogen Energy, 47(29), 13990-14007.

[18] Chen, K., Laghrouche, S., & Djerdir, A. (2021). Performance analysis of PEM fuel cell in mobile application under real traffic and environmental conditions. Energy Conversion and Management, 227. https://doi.org/ARTN11360210.1016/j.enconman.2020.113602

[19] Chen, P. Y., Chen, S. T., Hsu, C. S., & Chen, C. C. (2016). Modeling the global relationships among economic growth, energy consumption and CO2 emissions. Renewable & Sustainable Energy Reviews, 65, 420-431. https://doi.org/10.1016/j.rser.2016.06.074

[20] Council, W. E. (2021). Working Paper: Hydrogen On The Horizon: National Hydrogen Strategies. Retrieved 21.02.2022 from https://www.worldenergy.org/publications/entry/working-paper-hydrogen-on-the-horizon-national-hydrogen-strategies

[21] Cremers, C., Krausa, T. M. (2009). APPLICATIONS – PORTABLE | Military: Batteries and Fuel Cells. In Chemistry, Molecular Sciences and Chemical Engineering (pp. 13-21). Elsevier.

[22] Cunanan, C., Tran, M. K., Lee, Y., Kwok, S., Leung, V., & Fowler, M. (2021). A Review of Heavy-Duty Vehicle Powertrain Technologies: Diesel Engine Vehicles, Battery Electric Vehicles, and Hydrogen Fuel Cell Electric Vehicles. Clean Technologies, 3(2), 474-489. https://doi.org/10.3390/cleantechnol3020028

[23] D'Ovidio, G., Ometto, A., & Villante, C. (2020). A Novel Optimal Power Control for a City Transit Hybrid Bus Equipped with a Partitioned Hydrogen Fuel Cell Stack. Energies, 13(11). https://doi.org/ARTN268210.3390/en13112682

[24] Das, H. S., Tan, C. W., & Yatim, A. H. M. (2017). Fuel cell hybrid electric vehicles: A review on power conditioning units and topologies. Renewable & Sustainable Energy Reviews, 76, 268-291. https://doi.org/10.1016/j.rser.2017.03.056

[25] de Almeida, S. C. A., & Kruczan, R. (2021). Effects of drivetrain hybridization on fuel economy, performance and costs of a fuel cell hybrid electric vehicle. International Journal of Hydrogen Energy, 46(79), 39404-39414. https://doi.org/10.1016/j.ijhydene.2021.09.144

[26] de Miranda, P. E. V. (2019). Hydrogen Energy: Sustainable and Perennial. Science and Engineering of Hydrogen-Based Energy Technologies: Hydrogen Production and Practical Applications in Energy Generation, 1-38. https://doi.org/10.1016/B978-0-12-814251-6.00001-0

[27] Dixon, R. K., Li, J., Wang, M. Q. (2016). Progress in hydrogen energy infrastructure development—addressing technical and institutional barriers. In Compendium of Hydrogen Energy, Volume 2: Hydrogen

Storage, Distribution and Infrastructure focuses on the storage and transmission of hydrogen. (pp. 323-343). Woodhead Publishing.

[28] DMI hydrogen drones for Korean military. (2021). 6).

[29] Europe launches JIVE and MEHRLIN projects to deliver and refuel 144 fuel cell buses. (2017). 2).

[30] Folkesson, A., Andersson, C., Alvfors, P., Alakula, M., & Overgaard, L. (2003). Real life testing of a hybrid PEM fuel cell bus. Journal of Power Sources, 118(1-2), 349-357. https://doi.org/10.1016/S0378-7753(03)0 0086-7

[31] Fragiacomo, P., & Piraino, F. (2019). Fuel cell hybrid powertrains for use in Southern Italian railways. International Journal of Hydrogen Energy, 44(51), 27930-27946. https://doi.org/10.1016/j.ijhydene.2 019.09.005

[32] Gandia, L. M., Arzamendi, G., & Dieguez, P. M. (2013). Renewable Hydrogen Energy: An Overview. Renewable Hydrogen Technologies: Production, Purification, Storage, Applications and Safety, 1-17. https: //doi.org/10.1016/B978-0-444-56352-1.00001-5

[33] Gang, B. G., & Kwon, S. (2018). All-in-one portable electric power plant using proton exchange membrane fuel cells for mobile applications. International Journal of Hydrogen Energy, 43(12), 6331-6339. https://doi.org/10.1016/j.ijhydene.2018.02.006

[34] Goldi and Horizon plan fuel cell buses for new European project. (2019). 11).

[35] Graber, G., Calderaro, V., & Galdi, V. (2022). Two-Stage Optimization Method for Sizing Stack and Battery Modules of a Fuel Cell Vehicle Based on a Power Split Control. Electronics, 11(3). https://doi.org/AR TN36110.3390/electronics11030361

[36] Greene, D. L., Ogden, J. M., & Lin, Z. H. (2020). Challenges in the designing, planning and deployment of hydrogen refueling infrastructure for fuel cell electric vehicles. Etransportation, 6. https://doi.org/AR TN10008610.1016/j.etran.2020.100086

[37] Haghi, E., Shamsi, H., Dimitrov, S., Fowler, M., & Raahemifar, K. (2020). Assessing the potential of fuel cell-powered and battery-powered forklifts for reducing GHG emissions using clean surplus power; a game theory approach. International Journal of Hydrogen Energy, 45(59), 34532-34544. https://doi.org/10.1016/j.ijhydene.2 019.05.063

[38] Hienuki, S., Hirayama, Y., Shibutani, T., Sakamoto, J., Nakayama, J., & Miyake, A. (2019). How Knowledge about or Experience with Hydrogen Fueling Stations Improves Their Public Acceptance. Sustainability, 11(22). https://doi.org/ARTN633910.3390/su11226339

[39] Hu, G. P., Chen, C., Lu, H. T., Wu, Y., Liu, C. M., Tao, L. F., Men, Y. H., He, G. L., & Li, K. G. (2020). A Review of Technical Advances, Barriers, and Solutions in the Power to Hydrogen (P2H) Roadmap. Engineering, 6(12), 1364-1380. https://doi.org/10.1016/j.eng.2020.04.

[40] Inci, M., Buyuk, M., Demir, M. H., & Ilbey, G. (2021). A review and research on fuel cell electric vehicles: Topologies, power electronic converters, energy management methods, technical challenges, marketing and future aspects. Renewable & Sustainable Energy Reviews, 137. https://doi.org/ARTN11064810.1016/j.rser.2020.110648

[41] Intelligent Energy in project to extend flight time in military drones. (2021). 10).

[42] Ji, M. D., & Wang, J. L. (2021). Review and comparison of various hydrogen production methods based on costs and life cycle impact assessment indicators. International Journal of Hydrogen Energy, 46(78), 38612-38635. https://doi.org/10.1016/j.ijhydene.2021.09.142

[43] Jiang, S. Q., Suo, H. L., Zhang, T., Liao, C. Z., Wang, Y. X., Zhao, Q. L., & Lai, W. H. (2022). Recent Advances in Seawater Electrolysis. Catalysts, 12(2). https://doi.org/ARTN12310.3390/catal12020123

[44] Kar, S. K., Harichandan, S., Roy, B. (2022). Bibliometric analysis of the research on hydrogen economy: An analysis of current findings and roadmap ahead.

[45] Khani, N. G. (2022). Improving dynamic response of PEMFC using SMES and bidirectional DC/DC converter. Automatika, 63(4), 745-755. https://doi.org/10.1080/00051144.2022.2066768

[46] Kudiiarov, V., Lyu, J. Z., Semenov, O., Lider, A., Chaemchuen, S., & Verpoort, F. (2021). Prospects of hybrid materials composed of MOFs and hydride-forming metal nanoparticles for light-duty vehicle hydrogen storage. Applied Materials Today, 25. https://doi.org/ARTN101208 10.1016/j.apmt.2021.101208

[47] Lee, S., Kim, H. S., Park, J., Kang, B. M., Cho, C. H., Lim, H., & Won, W. (2021). Scenario-Based Techno-Economic Analysis of Steam Methane Reforming Process for Hydrogen Production. Applied Sciences-Basel, 11(13). https://doi.org/ARTN602110.3390/app11136 021

[48] Liu, Y. W., Li, Z. Y., Chen, Y. Z., & Zhao, K. G. (2020). A Novel Fuel-Cell Electric Articulated Vehicle and Its Drop-and-Pull Transport System. Energies, 13(14). https://doi.org/ARTN363210.3390/en13143632

[49] Lucia, U. (2014). Overview on fuel cells. Renewable & Sustainable Energy Reviews, 30, 164-169. https://doi.org/10.1016/j.rser.2013.09.025

[50] Luo, Y., Wu, Y. H., Li, B., Mo, T. D., Li, Y., Feng, S. P., Qu, J. K., & Chu, P. K. (2021). Development and application of fuel cells in the automobile industry. Journal of Energy Storage, 42. https://doi.org/ARTN10312410.1016/j.est.2021.103124

[51] Ma, S., Lin, M., Lin, T. E., Lan, T., Liao, X., Marechal, F., Van Herle, J., Yang, Y. P., Dong, C. Q., & Wang, L. G. (2021). Fuel cell-battery hybrid systems for mobility and off-grid applications: A review. Renewable & Sustainable Energy Reviews, 135. https://doi.org/ARTN11011910.1016/j.rser.2020.110119

[52] Maestre, V. M., Ortiz, A., & Ortiz, I. (2021). Challenges and prospects of renewable hydrogen-based strategies for full decarbonization of stationary power applications. Renewable & Sustainable Energy Reviews, 152. https://doi.org/ARTN11162810.1016/j.rser.2021.111628

[53] Maniam, K. K., Chetty, R., Thimmappa, R., & Paul, S. (2022). Progress in the Development of Electrodeposited Catalysts for Direct Liquid Fuel Cell Applications. Applied Sciences-Basel, 12(1). https://doi.org/ARTN50110.3390/app12010501

[54] Manoharan, Y., Hosseini, S. E., Butler, B., Alzhahrani, H., Fou, B. T., Ashuri, T., & Krohn, J. (2019). Hydrogen Fuel Cell Vehicles; Current Status and Future Prospect. Applied Sciences-Basel, 9(11). https://doi.org/ARTN229610.3390/app9112296

[55] Midilli, A., Kucuk, H., Topal, M. E., Akbulut, U., & Dincer, I. (2021). A comprehensive review on hydrogen production from coal gasification: Challenges and Opportunities. International Journal of Hydrogen Energy, 46(50), 25385-25412. https://doi.org/10.1016/j.ijhydene.2021.05.088

[56] Mishra, P., Krishnan, S., Rana, S., Singh, L., Sakinah, M., & Ab Wahid, Z. (2019). Outlook of fermentative hydrogen production techniques: An overview of dark, photo and integrated dark-photo fermentative approach to biomass. Energy Strategy Reviews, 24, 27-37. https://doi.org/10.1016/j.esr.2019.01.001

[57] Moradi, R., & Groth, K. M. (2019). Hydrogen storage and delivery: Review of the state of the art technologies and risk and reliability analysis. International Journal of Hydrogen Energy, 44(23), 12254-12269. https://doi.org/10.1016/j.ijhydene.2019.03.041
[58] NCMS funds project to integrate fuel cell into Nikola Reckless EV. (2019). 11).
[59] Nguyen, H.-U.-D., Nguyen, D-T., Taguchi, K. (2021). A portable soil microbial fuel cell for sensing soil water content. Measurement: Sensors, 18, Article 100231.
[60] Nicolay, S., Karpuk, S., Liu, Y. L., & Elham, A. (2021). Conceptual design and optimization of a general aviation aircraft with fuel cells and hydrogen. International Journal of Hydrogen Energy, 46(64), 32676-32694. https://doi.org/10.1016/j.ijhydene.2021.07.127
[61] Ogbonnaya, C., Abeykoon, C., Nasser, A., Turan, A., & Ume, C. S. (2021). Prospects of Integrated Photovoltaic-Fuel Cell Systems in a Hydrogen Economy: A Comprehensive Review. Energies, 14(20). https://doi.org/ARTN682710.3390/en14206827
[62] Olabi, A. G., Wilberforce, T., & Abdelkareem, M. A. (2021). Fuel cell application in the automotive industry and future perspective. Energy, 214. https://doi.org/ARTN11895510.1016/j.energy.2020.118955
[63] Olabi, A. G., Wilberforce, T., Sayed, E. T., Elsaid, K., & Abdelkareem, M. A. (2020). Prospects of Fuel Cell Combined Heat and Power Systems. Energies, 13(16). https://doi.org/ARTN410410.3390/en13164104
[64] Ong, B. C., Kamarudin, S. K., & Basri, S. (2017). Direct liquid fuel cells: A review. International Journal of Hydrogen Energy, 42(15), 10142-10157. https://doi.org/10.1016/j.ijhydene.2017.01.117
[65] Osmieri, L., & Meyer, Q. (2022). Recent advances in integrating platinum group metal-free catalysts in proton exchange membrane fuel cells. Current Opinion in Electrochemistry, 31. https://doi.org/ARTN31:10084710.1016/j.coelec.2021.100847
[66] Ozbek, E., Yalin, G., Karaoglan, M. U., Ekici, S., Colpan, C. O., & Karakoc, T. H. (2021). Architecture design and performance analysis of a hybrid hydrogen fuel cell system for unmanned aerial vehicle. International Journal of Hydrogen Energy, 46(30), 16453-16464. https://doi.org/10.1016/j.ijhydene.2020.12.2160360-3199/
[67] Ozcelep, Y., Bekdas, G., & Apak, S. (2021). Investigation of photovoltaic-hydrogen power system for a real house in Turkey: Hydrogen blending to natural gas effects on system design. International

Journal of Hydrogen Energy, 46(74), 36678-36686. https://doi.org/10.1016/j.ijhydene.2021.08.186

[68] Öztürk, A., Akay, R. G., Erkan, S., & Bayrakçeken Yurtcan, A. (2020). Introduction to fuel cells. In Direct Liquid Fuel Cells Fundamentals, Advances and Future (pp. 1-47). Academic Press.

[69] Pal, P., Ting, J-M., Agarwal, S., Ichikawa, T., Jain, A. (2021). The Catalytic Role of D-block Elements and Their Compounds for Improving Sorption Kinetics of Hydride Materials: A Review. Reactions, 2(3), 333-364.

[70] Palmer, C. (2022). Hydrogen-powered trains start to roll. Engineering.

[71] Parkhey, P., & Sahu, R. (2021). Microfluidic microbial fuel cells: Recent advancements and future prospects. International Journal of Hydrogen Energy, 46(4), 3105-3123. https://doi.org/10.1016/j.ijhydene.2020.07.019

[72] Peera, S. G., & Liu, C. (2022). Unconventional and scalable synthesis of non-precious metal electrocatalysts for practical proton exchange membrane and alkaline fuel cells: A solid-state co-ordination synthesis approach. Coordination Chemistry Reviews, 463. https://doi.org/ARTN21455410.1016/j.ccr.2022.214554

[73] Perna, A., Minutillo, M., Di Micco, S., & Jannelli, E. (2022). Design and Costs Analysis of Hydrogen Refuelling Stations Based on Different Hydrogen Sources and Plant Configurations. Energies, 15(2). https://doi.org/ARTN54110.3390/en15020541

[74] Pingkuo, L., Xue, H.,. (2022). Comparative analysis on similarities and differences of hydrogen energy development in the World's top 4 largest economies: A novel framework. International Journal of Hydrogen Energy, 47(16), 9485-9503.

[75] Pollet, B. G., Staffell, I., & Shang, J. L. (2012). Current status of hybrid, battery and fuel cell electric vehicles: From electrochemistry to market prospects. Electrochimica Acta, 84, 235-249. https://doi.org/10.1016/j.electacta.2012.03.172

[76] Psoma, A., & Sattler, G. (2002). Fuel cell systems for submarines: from the first idea to serial production. Journal of Power Sources, 106(1-2), 381-383. https://doi.org/PiiS0378-7753(01)01044-8Doi10.1016/S0378-7753(01)01044-8

[77] Randive, V., Subramanian, S. C., & Thondiyath, A. (2021). Design and analysis of a hybrid electric powertrain for military tracked vehicles. Energy, 229. https://doi.org/ARTN12076810.1016/j.energy.2021.120768

[78] Rodriguez, J., & Amores, E. (2020). CFD Modeling and Experimental Validation of an Alkaline Water Electrolysis Cell for Hydrogen Production. Processes, 8(12). https://doi.org/ARTN163410.3390/pr8121634

[79] Sagaria, S., Neto, R. C., & Baptista, P. (2021). Assessing the performance of vehicles powered by battery, fuel cell and ultra-capacitor: Application to light-duty vehicles and buses. Energy Conversion and Management, 229. https://doi.org/ARTN11376710.1016/j.enconman.2020.113767

[80] Strasser, K. (2010). H2/O2-PEM-fuel cell module for an air independent propulsion system in a submarine. In Handbook of Fuel Cells. Wiley.

[81] Sürer, M. G., Arat, H. T. (2022). Advancements and current technologies on hydrogen fuel cell applications for marine vehicles. International Journal of Hydrogen Energy.

[82] Tanc, B., Arat, H. T., Baltacioglu, E., & Aydin, K. (2019). Overview of the next quarter century vision of hydrogen fuel cell electric vehicles. International Journal of Hydrogen Energy, 44(20), 10120-10128. https://doi.org/10.1016/j.ijhydene.2018.10.112

[83] Tarkowski, R., Uliasz-Misiak, B. (2022). Towards underground hydrogen storage: A review of barriers. Renewable and Sustainable Energy Reviews, 162, Article 112451.

[84] Thomas, C. E. S. (2009). Transportation options in a carbon-constrained world: Hybrids, plug-in hybrids, biofuels, fuel cell electric vehicles, and battery electric vehicles. International Journal of Hydrogen Energy, 34(23), 9279-9296. https://doi.org/10.1016/j.ijhydene.2009.09.058

[85] Tian, Z., Lv, H., Zhou, W., Zhang, C. M., & He, P. F. (2022). Review on equipment configuration and operation process optimization of hydrogen refueling station. International Journal of Hydrogen Energy, 47(5), 3033-3053. https://doi.org/10.1016/j.ijhydene.2021.10.238

[86] Tong, X., Dai, H. C., Lu, P. T., Zhang, A. Y., & Ma, T. (2022). Saving global platinum demand while achieving carbon neutrality in the passenger transport sector: linking material flow analysis with integrated assessment model. Resources Conservation and Recycling, 179. https://doi.org/ARTN10611010.1016/j.resconrec.2021.106110

[87] Usai, L., Hung, C. R., Vasquez, F., Windsheimer, M., Burheim, O. S., & Stromman, A. H. (2021). Life cycle assessment of fuel cell systems for light duty vehicles, current state-of-the-art and future impacts. Journal of Cleaner Production, 280. https://doi.org/ARTN12508610.1016/j.jclepro.2020.125086

[88] Vaghari, H., Jafarizadeh-Malmiri, H., Berenjian, A., Anarjan, N. (2013). Recent advances in application of chitosan in fuel cells. Sustainable Chemical Processes, 1, Article 16.

[89] Vinodh, R., Atchudan, R., Kim, H-J., Yi, M. (2022). Recent Advancements in Polysulfone Based Membranes for Fuel Cell (PEMFCs, DMFCs and AMFCs) Applications: A Critical Review. Polymers, 14(2).

[90] Volkart, K., Densing, M., De Miglio, R., Priem, T., Pye, S., & Cox, B. (2017). The Role of Fuel Cells and Hydrogen in Stationary Applications. Europe's Energy Transition: Insights for Policy Making, 189-205. https://doi.org/10.1016/B978-0-12-809806-6.00023-7

[91] Wang, J., Jia, R., Liang, J., She, C., & Xu, Y. P. (2021). Evaluation of a small drone performance using fuel cell and battery; Constraint and mission analyzes. Energy Reports, 7, 9108-9121. https://doi.org/10.101 6/j.egyr.2021.11.225

[92] Wang, J. Y., Wang, H. L., & Fan, Y. (2018). Techno-Economic Challenges of Fuel Cell Commercialization. Engineering, 4(3), 352-360. https://doi.org/10.1016/j.eng.2018.05.007

[93] Wilberforce, T., Alaswad, A., Palumbo, A., Dassisti, M., & Olabi, A. G. (2016). Advances in stationary and portable fuel cell applications. International Journal of Hydrogen Energy, 41(37), 16509-16522. https: //doi.org/10.1016/j.ijhydene.2016.02.057

[94] Wu, W., Partridge, J. S., & Bucknall, R. W. G. (2018). Simulation of a stabilised control strategy for PEM fuel cell and supercapacitor hybrid propulsion system for a city bus. International Journal of Hydrogen Energy, 43(42), 19763-19777.https://doi.org/10.1016/j.ijhydene.201 8.09.004

[95] Xun, D. Y., Sun, X., Geng, J. X., Liu, Z. W., Zhao, F. Q., & Hao, H. (2021). Mapping global fuel cell vehicle industry chain and assessing potential supply risks. International Journal of Hydrogen Energy, 46(29), 15097-15109. https://doi.org/10.1016/j.ijhydene.2021.02.041

[96] Yodwong, B., Guilbert, D., Phattanasak, M., Kaewmanee, W., Hinaje, M., & Vitale, G. (2020). AC-DC Converters for Electrolyzer Applications: State of the Art and Future Challenges. Electronics, 9(6). https: //doi.org/ARTN91210.3390/electronics9060912

[97] Yue, M. L., Jemei, S., Zerhouni, N., & Gouriveau, R. (2021). Proton exchange membrane fuel cell system prognostics and decision-making: Current status and perspectives. Renewable Energy, 179, 2277-2294. ht tps://doi.org/10.1016/j.renene.2021.08.045

[98] Zandalinas, S. I., Fritschi, F. B., & Mittler, R. (2021). Global Warming, Climate Change, and Environmental Pollution: Recipe for a Multifactorial Stress Combination Disaster. Trends in Plant Science, 26(6), 588-599. https://doi.org/10.1016/j.tplants.2021.02.011

[99] Zhang, B., Zhang, S-X., Yao, R., Wu, Y-H., Qiu, J-S., . (2021). Progress and prospects of hydrogen production: Opportunities and challenges. Journal of Electronic Science and Technology, 19(2), Article 100080.

5

Hydrogen Production and Bunkering from Offshore Wind Power Plants for Green and Sustainable Shipping

Sabri Alkan

Department of Motor Vehicles and Transportation Technologies, Maritime Vocational School of Higher Education, Bandırma Onyedi Eylül University, Turkey
E-mail: salkan@bandirma.edu.tr

Abstract

Over the past two decades, greenhouse gas (GHG) emissions have drawn much attention as they play a significant role in environmental concerns in the shipping industry and many others. Some possible ways to cut down on global GHG emissions are natural gas, hydrogen, synthetic fuels, and renewable energy. Hydrogen is essential to a sustainable energy solution because it can store energy and be used to produce fuels like ammonia and methanol. Furthermore, if hydrogen production is combined with renewables, it can meet the zero-emission target. The primary source of hydrogen is water, which is found in large amounts offshore. Thus, countries are looking into producing hydrogen from offshore wind energy. On the other hand, the maritime industry is looking very closely at hydrogen for ships' propulsion and auxiliary power needs so that long-distance shipping can be done with no emissions. But the biggest problem with reaching this goal is that there is no cheap way to store a lot of hydrogen on board. This chapter examines producing hydrogen at offshore wind power plants and green hydrogen refueling ships for long-distance, environmentally friendly shipping.

Keywords: Offshore wind energy, green hydrogen, refueling, ships, synthetic fuels, fuel cell

5.1 Introduction

The demand for energy worldwide is changing quickly, and there are concerns about the security of energy supplies, rising fuel prices, and how emissions affect global climate change. Since carbon-based energy sources are used so much, fewer fossil-based energy sources are left, which causes security, price, and availability problems. Moreover, greenhouse gas (GHG) emissions have reached a critical level and raised environmental concerns in the last two decades. Due to fossil fuels' environmental damage, people are looking for clean energy sources, such as renewable energy (World Bank, 2019; Pavithran et, al., 2021). For example, natural gas, hydrogen, renewable sources, and synthetic fuels have all been suggested as ways to reduce GHG emissions. However, delivering energy from renewable clean energy sources such as wind, waves, and solar is impossible when and where it is needed. So, irregular and unpredictable energy production means excess alternative energy must be temporarily stored.

Hydrogen is known worldwide as a way to store extra energy from renewable sources for a long time without wasting much (Grzegorz Pawelec, 2021). It can be used in almost every sector where energy is needed, such as transportation, homes, services, and industry, which offers significant advantages. Since the only thing that comes out of using hydrogen is water vapor, it is almost completely free of CO_2, acid gases, and other pollutants. Hydrogen can be obtained from several sources, such as fossil fuels, nuclear power, and renewable sources. Improving the hydrogen economy depends on making hydrogen production cheaper, more efficient, and better for the environment. With offshore solar, wave, and wind energy facilities, it may be possible to make hydrogen in a way that is good for the environment. So, after being compressed, stored in tanks, or moved to land through pipelines (Cecchinato et al., 2021; Stori, 2021), hydrogen can power cars, factories, and power grids.

Because of environmental concerns, there will be more demand for hydrogen in the next few years (DNV, 2021a). Steam-reforming natural gas and coal gasification technologies currently produce hydrogen based on fossil fuels. On the other hand, the electrolysis method is used in some specific applications where a small amount of pure hydrogen is required. Hydrogen can compensate for the intermittent production of renewable energy sources since it can be stored and sustainably produced from wind power through water electrolysis. With these qualities, hydrogen is becoming one of the most critical energy carriers to solve the climate change problem and reach the Paris Agreement's goals (DNV, 2021a).

Many industries need help reducing their share of global greenhouse gas emissions. The transport sector is one of them, being responsible for a significant part of the emissions (23%) (International Energy Agency, 2020). Therefore, the transport sector needs to make more efforts to reduce GHG emissions. Transportation modes like land, maritime, rail, and airways share this rate. Although maritime transportation is the most energy-efficient mode of transportation in terms of energy per ton kilometer, it is responsible for almost 2%–3% of CO_2 emissions worldwide (IMO, 2020). Nearly 3% of the world's final energy demand is presently consumed by ships, mainly by international maritime transportation (DNV-GL, 2019; DNV, 2021a; DNV GL-Maritime, 2019). Even though this rate seems low, the gross domestic product (GDP) is expected to double by 2050 (British Petroleum, 2020). This means that more cargo ton-miles will be moved by sea. Thus, ton-miles in cargo transportation, with a total growth of 28% between 2019 and 2035, are expected to increase in almost all shipping categories (IMO, 2020).

In 2020, the International Maritime Organization's (IMO) global sulfur cap regulation in maritime transportation to decrease emissions dramatically changed ship fuel types (DNV, 2021b). The major shift has been toward lighter distillates or other less sulfurous fuels. However, much marine heavy fuel oil is still used on ships where scrubbers are installed. The IMO's long-term goal is an absolute 50% reduction in CO_2 emissions from 2008 to 2050 (DNV, 2021b; IMO, 2022, 2018), with the support of both shipowners and governments. The transition from oil to gas and ammonia, followed by other low- and zero-carbon fuel alternatives, will satisfy this plan (Smith et al., 2021). Cutting emissions in shipping would have a favorable impact on the environment.

Today, the fuel mix transition in maritime transportation, dominated by almost all oil, is changing to a mix dominated by low- and zero-carbon fuels (42%) and natural gas (39%, mostly liquefied natural gas) in 2050 (DNV, 2021a, 2021b; Smith et al., 2021). Low-carbon fuels include ammonia, hydrogen, and other electro-based fuels like e-methanol. The pressure of decarbonization will cause a significant change in the fuel mix in the marine industry over the coming decades (DNV, 2021b). Compared to road transport, the maritime sector's electrification potential is limited to short-sea transport. This is because the energy density of batteries is likely to stay too low for them to play a more significant role in ocean transport, both now and in the future. There is no essential battery-electric way to reduce carbon emissions in long-distance shipping (DNV, 2021b; Jackson et al., 2020). Even though electrification is the future of environmentally friendly transportation, it is much more challenging to use in the maritime industry. Instead, relying on alternative fuels for large carriers may give quick results in the short term, but hydrogen is a crucial part of reaching the net-zero goal in the long term.

Reaching the goal of zero emissions for long-distance shipping with ships that run on batteries is the hardest because of the amount of energy they can hold. Also, a critical bunkering problem exists for hydrogen and battery-powered ships traveling long distances. However, if hydrogen is combined with offshore renewable energy sources, it can solve the bunkering problem and give ocean-going ships a goal of zero emissions. With hydrogen-based fuels and fuel cells, offshore wind power plants can power and refuel ships in a way that does not release any pollution. Ultimately, ships that use the electricity and hydrogen made by offshore wind power plants can reduce energy transfer costs and pave the way for green shipping.

5.2 Offshore Wind Power Technology

Since Sweden launched its first 300-kW wind turbine in 1990, offshore wind technology has come a long way, and costs have also gone down, making electric power generation more competitive. Today, 116 offshore wind farms with 25 GW of capacity meet 3% of European electricity consumption, and this amount is expected to exceed 110 GW in 2030 and over 400 GW in 2050 (Cecchinato et al., 2021; Lee and Zhao, 2021).

Offshore wind farms can significantly affect the energy available, especially in populated coastal areas. Also, offshore wind power, which started in shallow waters, is now making its way into deeper waters, such as the oil and gas industry. The wind blows faster and more smoothly on offshore sites than on land, making faster and more stable wind energy production possible. The increased wind speed with distance from the shore makes reaching excellent wind fields at reasonable distances from the load centers possible. High and continuous wind power means more electricity is produced per turbine.

Offshore wind plants are much more complicated and take a lot more time to set up, run, and maintain than onshore plants (WindEurope, 2021; World Bank, 2019). This is because they have to do a lot of hard things during installation, operation, and maintenance. Figure 5.1 shows an overview of critical offshore wind activities. Also, the offshore wind energy industry needs onshore and offshore logistics activities (DNV, 2022; WindEurope, 2021). One of the most significant barriers to supporting offshore wind projects is transferring technicians in and out of turbines and substations

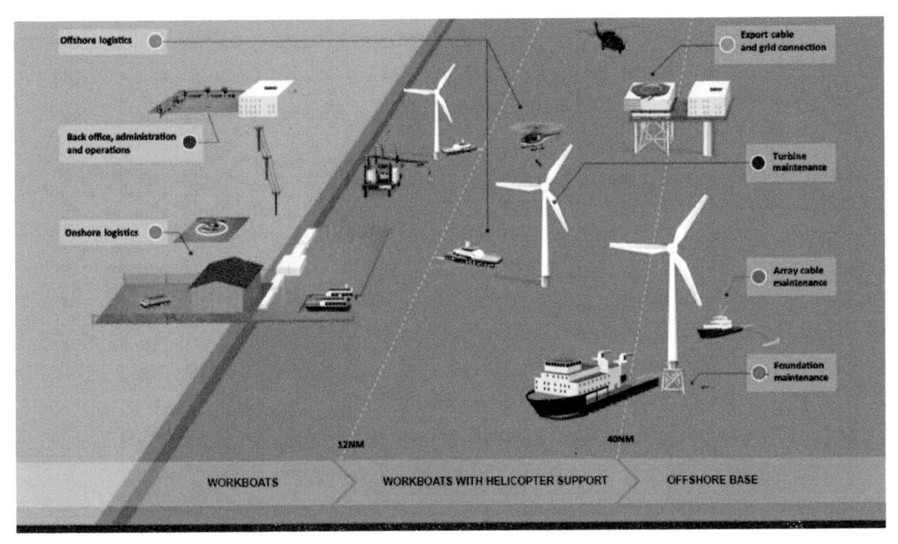

Figure 5.1 Overview of critical offshore wind O&M activities (GL Garrad Hassan, 2013).

to run the job. Transport time and accessibility are the two most important factors. The distance between offshore wind power plants and the base where they are run and maintained increases the time it takes for teams to get to the field and the chance of getting tired. This means that teams have less time to do active work. How long it takes to get safely to a turbine from a ship depends on its speed and the weather at sea. These factors depend more on the average sea conditions at a particular place than on how long they take. Accessibility is essential, especially for unplanned maintenance, since the project operator usually cannot plan breaks in production for times when sea conditions are better. When planning the O&M approach for a project, the owner will try to lower the total cost (direct cost plus loss of production) by looking for ways to reduce travel time and make reaching turbines easier.

The complexity of the bases or substructures of the turbines is the biggest technical challenge to the growth and development of offshore wind energy. The cost and feasibility of offshore wind power plants depend on the depth of the water, how far away they are from the shore, and how the seabed condition is (Jiang, 2021; Sumer and Kirca, 2022). Increasing water depth, the complexity of the seabed conditions, and the distance from shore all increase the costs of offshore foundations due to their additional complexities. Technological developments are still required for economically competitive wind energy over deeper water sites. The oil and gas industry has developed much technology, and now the wind industry is working hard to obtain that technology. Offshore wind turbines have either a fixed substructure up to 60 meters or a floating substructure in deeper water.

5.2.1 Fixed offshore wind turbines

Fixed offshore wind turbines dominate the current offshore wind energy market. Foundation types for bottom-fixed turbines can be classified as monopile, gravity base, and jacket. Figure 5.2 illustrates the most common fixed offshore wind turbine concepts (Jiang, 2021). Monopiles have a minimal footprint on the seafloor and are preferred at shallow depths, and their simple design provides minimal conditions to transition from land to open seas. Monopile applications are depth-limited up to 60 m because of their inherent flexibility. So, the main dimensions of the monopiles, like length, diameter, and thickness, need to get bigger to accommodate deeper water (Jiang, 2021). Keeping a monopile stiff enough means spending more money depending on how much deeper the water gets.

On the other hand, as depth increases, installation methods and tools like pile hammers and lifting vessels may need to be more specialized and cost more (Jiang, 2021). For monopile applications, the ideal limits are up to 30 m. Beyond 30 m of water depth, transitional depth technology is crucial to accessing the floating systems and the entire offshore wind resource. When

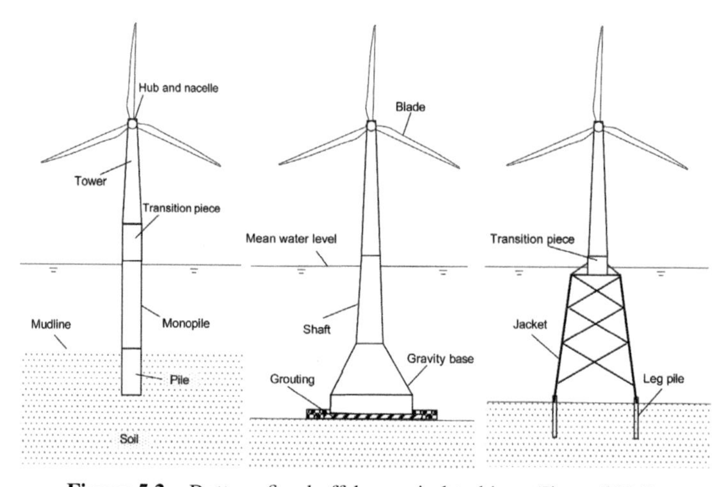

Figure 5.2 Bottom-fixed offshore wind turbines (Jiang, 2021).

the wind energy resources at the transition depths (30–60 m deep) for Class 5 wind and above are evaluated, it is seen that there is a potential of over 250 GW globally (Musial et al., 2006). Many wind turbines have been put up in transitional water depths off the coast of Europe to take advantage of this potential. However, because floating systems can be deployed anywhere, they will soon be able to replace fixed-base foundations, thanks to mass production and a lot of wind turbine innovation.

5.2.2 Floating offshore wind turbines

About 80% of the offshore wind resource is in waters deeper than 60 m, where wind turbines cannot be fixed to the seabed. At water depths greater than 60 m, a floating foundation may be the optimal choice. The buoyancy of the floating substructure is needed to keep the weight of the turbine steady and to control its pitch, roll, and heave movements. Even though the offshore oil and gas industry is an inspiration for floating wind turbine technology, there are differences in how they are built, installed, and run. On an oil rig, the payload and wave loads are critical. However, because of how floating wind turbines are loaded, the overturning moments caused by the wind determine how they are built and how they work. Floating offshore wind turbines can be created in many different ways due to the anchors, mooring systems, buoyancy tanks, and ballast configurations. However, designers have to think about factors that could go wrong and cause the cost of the system to be too high for wind applications. Even though it is possible to design a floating platform system that most designers can agree on, reaching the best design will not be possible because of how nature works. Over the past 10 years, demonstration projects have focused on improving the design of floating infrastructure, mooring systems, and mass production since these are the most critical parts of floating wind turbines that help keep costs down. Even though there are more than 40 floating wind demonstration projects, the industry has yet to agree on a single design (DNV, 2022). Figure 5.3 shows platform architectures assessed for floating offshore wind turbine foundations. Floating platforms are spar, semi-submersible, and tension leg (TLP) (Jiang, 2021).

5.2.3 Offshore substations

Offshore substations are critical in transforming the generated electricity and transmitting it to the planned location by high-voltage export cables

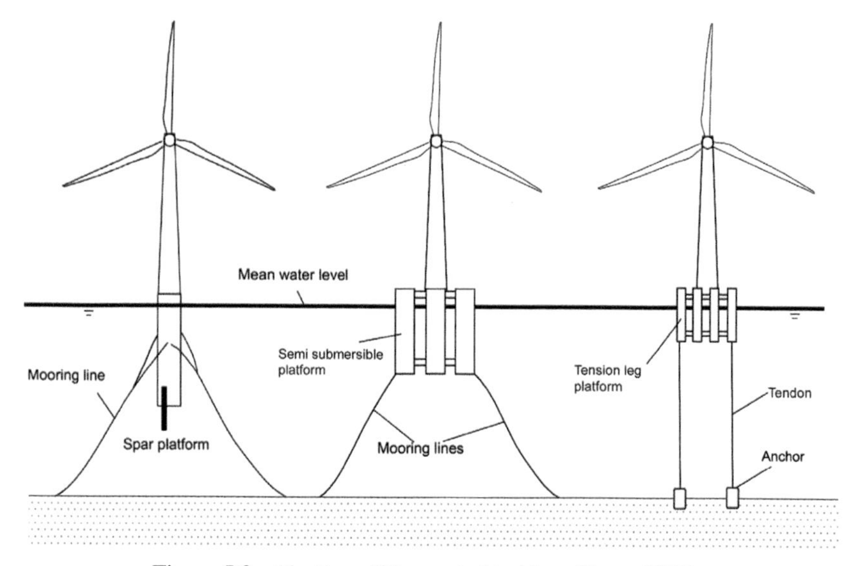

Figure 5.3 Floating offshore wind turbines (Jiang, 2021).

(Figure 5.4). Turbines are connected to offshore substations by array cables. Offshore substations, export, and array cables can be severely exposed to the dynamic conditions of the marine environment. So, they must be carefully watched during the design, installation, and use phases (European Union, 2016; GL Garrad Hassan, 2013). With the move to deeper waters, it is clear that floating offshore wind farms will need floating substations. Developing dynamic cables with higher voltage and power levels than those currently available and electrical hardware like transformers and switchgear will make these goals possible. Moreover, floating substations could lead to hydrogen production, storage, and transfer in the following decades.

Typically, wind power plants more distant than 10 km from land have offshore substations. The substation accommodates the transformers needed to raise the distribution voltage of the inter-array cables (33 kV or above) to a higher voltage of 110–245 kV. The export cables then transmit the power from the offshore substation to the land. There is a necessity for advanced electrical equipment ratings and more extensive substations as wind farm capacities grow and move farther offshore. The losses in the electrical system can become significant when wind power plants are located at considerable distances from shore. Voltages are stepped up to minimize losses, such as from 33 to 115 kV (Leanwind, 2015).

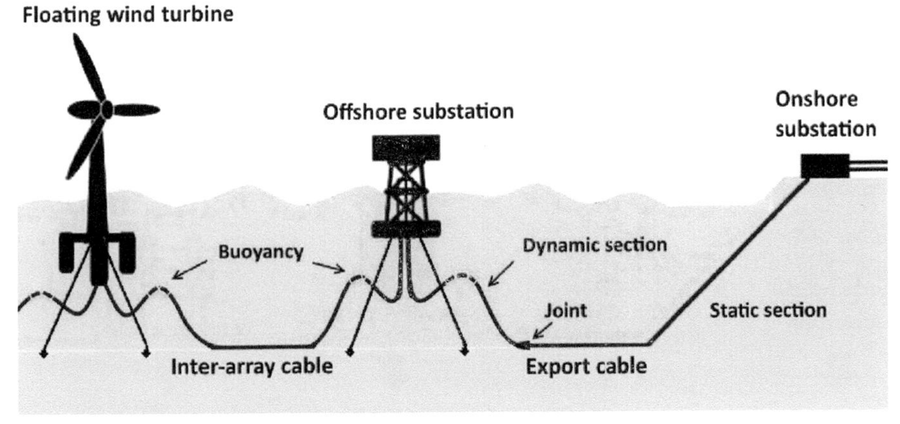

Figure 5.4 Offshore substation and cable connections (Lerch et al., 2021).

5.3 Green Hydrogen Production from Offshore Wind

Hydrogen is one of the most common elements in the universe, but it is only found in large amounts on Earth if combined with other elements. In general, it can combine with oxygen to form water; by combining with carbon, it can form coal, oil, natural gas, and biomass. Hydrogen has been provided for different industrial purposes in the last century and is extensively used, especially in the chemical and petroleum industries. Climate change, energy security concerns, and hydrogen's ability to transport energy have made it an essential subject of energy research and policy worldwide. However, the change to a hydrogen economy faces many technical, economic, and social challenges regarding production, distribution, storage infrastructure, conversion, and end-use (Cecchinato et al., 2021).

Hydrogen can be obtained from both fossil and non-fossil resources, like water, using either electricity or heat. While there are different ways to produce hydrogen commercially, some methods are already in use, while others are under development. Gasification of coal, steam reforming of natural gas, reforming of natural gas with carbon capture, and electrolysis of water (Figure 5.5) are four promising ways to produce hydrogen. The four most critical industrial processes for producing hydrogen are brown, gray, blue, and green, depending on how much pollution they reduce. The main criticism of the production methods made from fossil sources is that they cause emissions. To achieve a net-zero emission target, renewable energy sources such as solar, wave, and wind are essential for generating hydrogen (ABS, 2020).

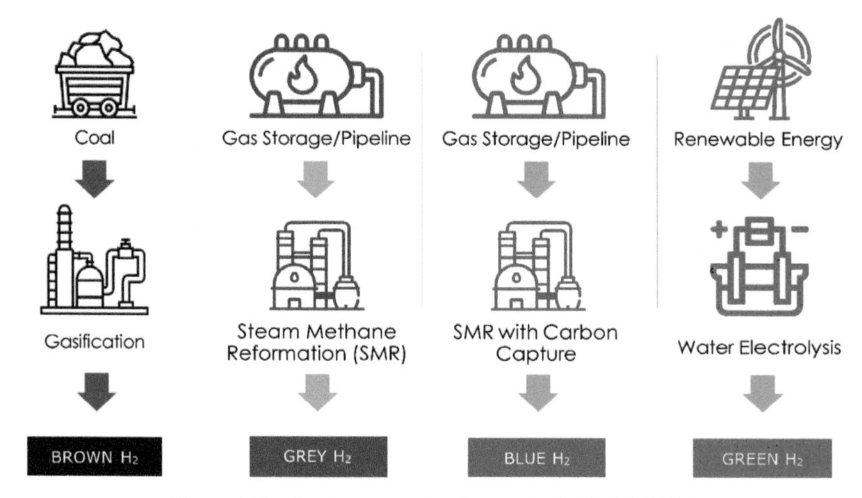

Figure 5.5 Hydrogen production methods (ABS, 2020).

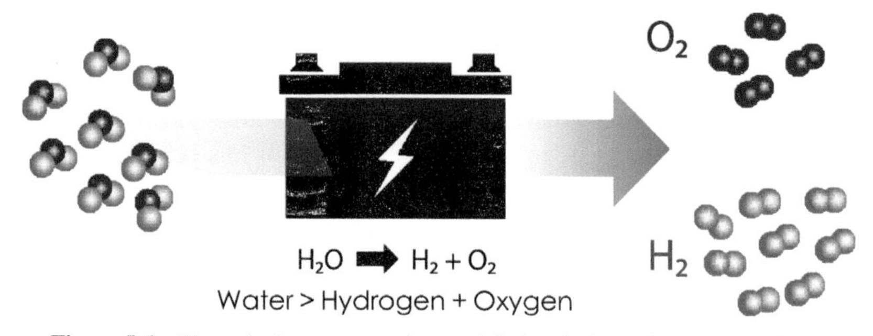

Figure 5.6 Electrolysis process and essential chemical reactions (ABS, 2020).

In the electrolysis process, a direct electric current is applied to the electrolyte, causing chemical reactions at the electrodes that break down reactants. During water electrolysis, water molecules break down into their two main parts, hydrogen and oxygen, without giving off any extra emissions. An electrolyzer carries out the electrolysis process, and Figure 5.6 shows the main chemical reactions of electrolysis.

Due to the unpredictable nature of alternative energy sources, making energy continuously from them is impossible. The most promising way to solve this problem is to obtain offshore hydrogen from renewable energy sources. Electrolysis can be used to produce hydrogen at several different scales offshore. The hydrogen can then be compressed, stored, or moved in tanks or pipelines (Calado and Castro, 2021; DNV, 2021a). Lastly, hydrogen can be used by power vehicles, factories, or power plants to obtain steam or electricity.

Solar and wind power are two crucial commercially available technologies to provide electricity for water electrolysis and achieve the net-zero emissions target. The ability to produce hydrogen from solar and wind energy by electrolysis will be cost-effective compared to other commercial methods, depending on the further development of electrolyzers and widespread hydrogen production on a large scale until 2030 (Lee and Zhao, 2021; Wind Energy Ireland, GreenTech Skillnet, 2022). It is noticeable that the hydrogen economy is very close to catching this target, considering the developments in recent years.

Due to its high wind speeds and potential, offshore wind power is becoming more attractive and competitive for less intermittent electricity and hydrogen production. Over the next 50 years, offshore wind technologies will benefit from large-scale hydrogen production (ABS, 2020; Stori, 2021; World Bank, 2019). Using hydrogen as an energy carrier in offshore power generation requires considering options for the location and design of conversion facilities (Stori, 2021). An offshore-based facility is needed to produce hydrogen from offshore wind energy, which can be transferred onshore around the offshore power generation facility (Calado and Castro, 2021). It should be noted that offshore-based hydrogen production requires supplementary construction in the marine environment and equipment that can be used in the long term. Also, people who work at offshore production facilities have to travel and do their jobs in the sea (European Union, 2016).

Hydrogen production from offshore wind energy has two main pathways: onshore and offshore. Most stakeholders think about how well and how much each method of making hydrogen works and how much it costs. In the first case, hydrogen is made by electrolysis on land, assuming electricity from an offshore wind power plant is used to produce hydrogen. The electricity produced by offshore wind turbines is collected in a substation and transmitted via high-voltage electrical cables to the onshore substation. Hydrogen is then stored via the electrolysis carried out on land (Figure 5.7). Hydrogen production at an onshore location close to the offshore wind power plant would help a land-based power grid in several ways. So, when conventional sources in the onshore grid are not enough to meet power needs, electricity from offshore sources can be used to produce hydrogen. In the second case, the method is classified as centralized and decentralized based on where the electrolysis process is done offshore. Hydrogen and electrical connections must be consolidated between each platform and the unit that

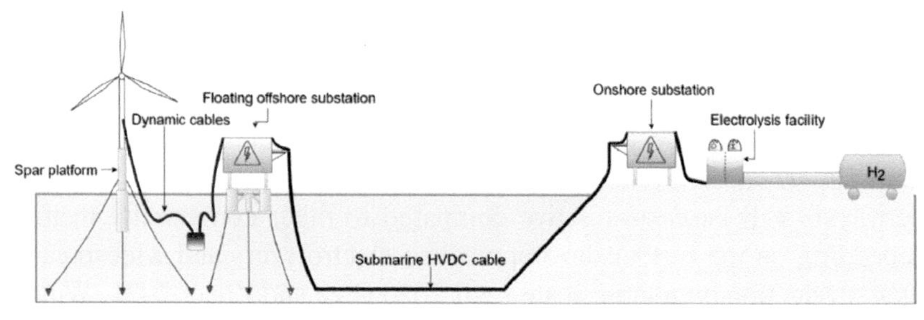

Figure 5.7 Onshore electrolysis coupled with offshore wind energy (Ibrahim et al., 2022).

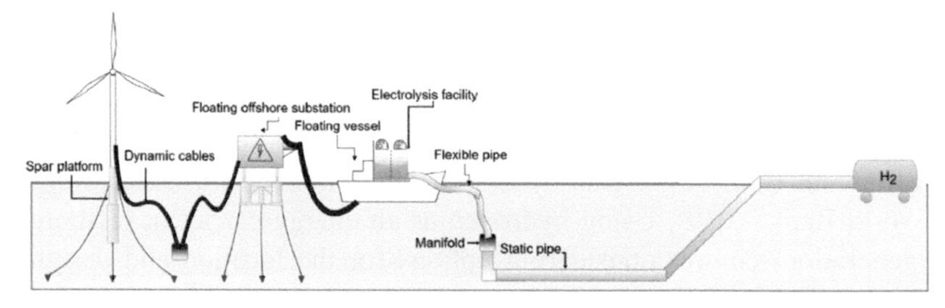

Figure 5.8 Centralized offshore electrolysis on a floating vessel (Ibrahim et al., 2022).

produces hydrogen in the centralized system. Hydrogen produced offshore can be delivered to onshore facilities in various ways. This typology has an offshore power plant that produces hydrogen, which is then sent to land for storage and distribution.

The offshore electrolysis facility is installed in a centralized configuration on a floating vessel. This system makes it easier and faster to get to specific turbines for maintenance and is usually less complicated than the decentralized one. A suggested arrangement of the system is shown in Figure 5.8. The term "decentralized" means conducting electrolysis on each turbine structure. The decentralized typology comprises electrolyzers, cooling units, seawater desalination units, hydrogen buffers, and a battery system for backup power, similar to the centralized typology. The decentralized system plans to use a submarine manifold to collect the flexible pipes from all the turbines. The hydrogen is then sent to land through a more extensive subsea pipeline system that stays in place (Figure 5.9).

There have been different demonstration projects on hydrogen production with offshore wind energy in the last decade. The Dolphyn project is one of the projects that include innovative solutions for decentralized hydrogen production (Caine et al., 2021; ERM, 2019). In the Dolphyn project, a

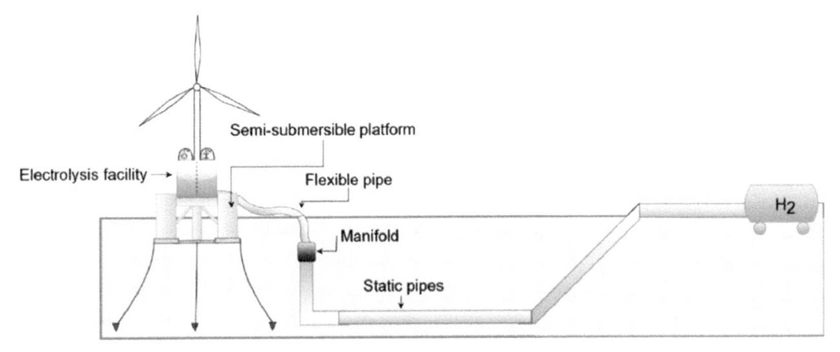

Figure 5.9 Decentralized offshore electrolysis on a floating wind turbine substructure (Ibrahim et al., 2022).

semi-submersible floating wind turbine substructure provides enough deck space for seawater desalination, hydrogen production, and storage facilities (Figure 5.10). Floating wind turbines create a powerful option to combine the high capacity factor of offshore wind and green hydrogen.

Large-scale offshore hydrogen production requires robust transport methods. The highest potentials are transported in gaseous form, in liquid form, and after incorporation into a solid or liquid hydrogen carrier. For economic reasons, hydrogen must be compressed before it can be transported to land in gaseous form. The shipping may be done via a pipeline between an offshore production facility and an onshore receiving facility. Storing or transporting hydrogen with current technology involves some difficulties. Storage of hydrogen is an important goal in developing the hydrogen economy. Due to the high pressures and cryogenic storage conditions generally, focus on chemical compounds release hydrogen reversibly upon heating. Storing hydrogen lightly and compactly for mobile applications forms the spine of hydrogen research. Although hydrogen has a good energy density by weight in its gaseous form, it has a low energy density by volume compared to hydrocarbons, which is also a disadvantage. Therefore, larger volumes are required to store hydrogen. Increased gas pressure will improve energy density by volume but require smaller but not lighter container tanks. However, applying more power to the compressor to compress the hydrogen will result in higher compression and more energy loss with each compression step.

On the other hand, liquid hydrogen requires cryogenic storage conditions due to its boiling point of around 20,268 K (–252.882 °C). Therefore, the liquefaction process also results in a significant energy loss. Also, tanks need to be well insulated; so they do not boil, which makes them expensive and fragile.

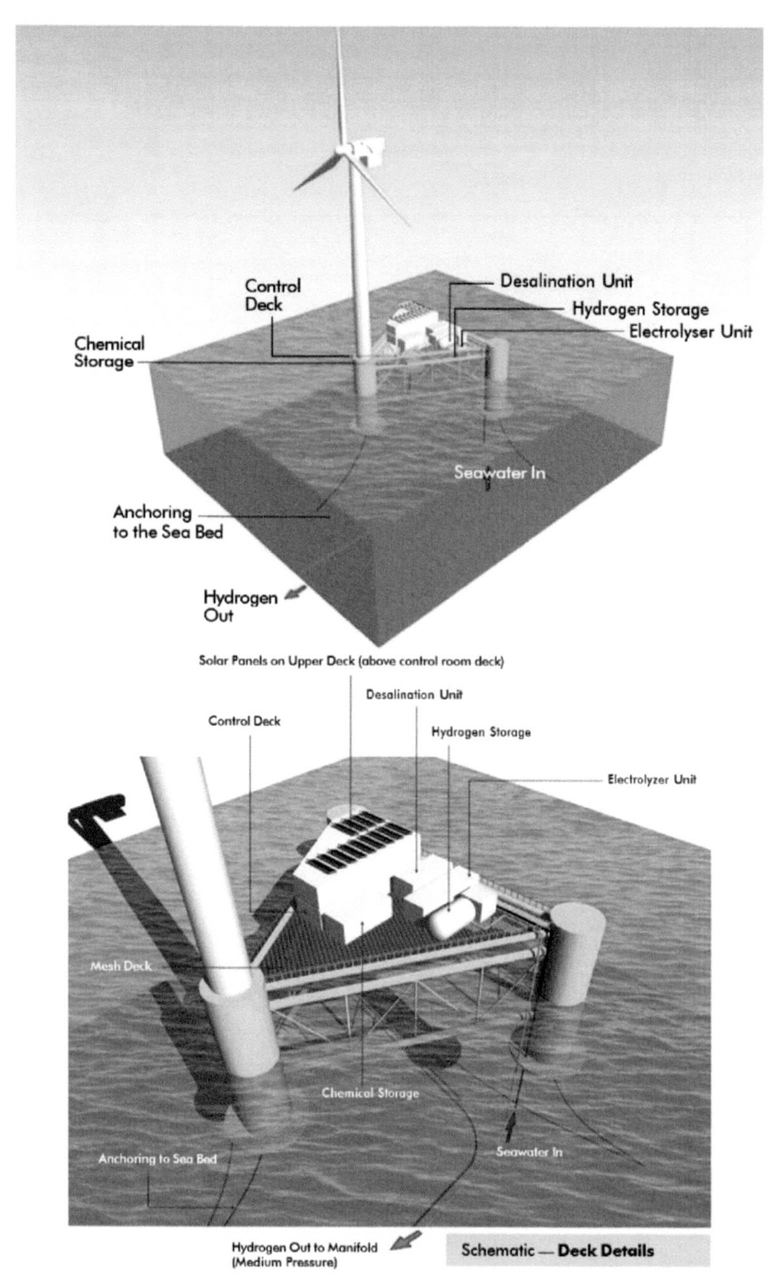

Figure 5.10 Decentralized offshore hydrogen production in the Dolphyn project (Caine et al., 2021; ERM, 2019)

5.4 Hydrogen Power for Ships

What fuel will power ships in the short and long term? (Grzegorz Pawelec, 2021). This is a fundamental question for the maritime industry. Although it is tricky to answer this question today, the answer is green hydrogen and green hydrogen-based synthetic fuels in the long run. In the current technology, synthetic fuels, biofuels, methanol, ammonia, and hydrogen are the low- or zero-carbon fuel choices (DNV, 2021b, 2021a). Many problems, such as safety, regulations, weight, space, and cost, make storing much hydrogen onboard hard. These factors have made people want to learn more about other hydrogen-based energy carriers, like ammonia. Hydrogen as a direct fuel or hydrogen-based fuels such as ammonia needs green hydrogen production to achieve the net-zero emissions goal by 2050 (DNV, 2021b). Figure 5.11 shows the capability of hydrogen to generate synthetic fuels for ships.

Hydrogen is the leader of several clean fuel alternatives currently being piloted. Nearly half of the global projects that aim for zero emissions focus on hydrogen as a low-carbon fuel source (Fahnestock and Bingham, 2021). One advantage of hydrogen over other fuels is that it can be used by installing fuel cells on ships already in use. Different countries have tested hydrogen-powered ferries and smaller transports, but they still need to be tested by an extensive shipping company (Fahnestock and Bingham, 2021).

With zero-emission fuel cells, hydrogen is an exciting energy source for shipowners who want to reduce their fleets' environmental impact and keep up with changing rules about sustainability. Hydrogen can help decarbonize shipping as an alternative fuel and clean electricity source. With hydrogen as a fuel, ships can run their propulsion and auxiliary power systems without putting out GHGs. While the direct use of hydrogen as a ship fuel with fuel cells is the most energy-efficient choice, it is also possible to use it as a component to produce synthetic e-fuels. Hydrogen production from fossil fuels must be replaced with renewable or low-carbon alternatives. When steam methane reforming (SMR) is used to produce and use hydrogen, it

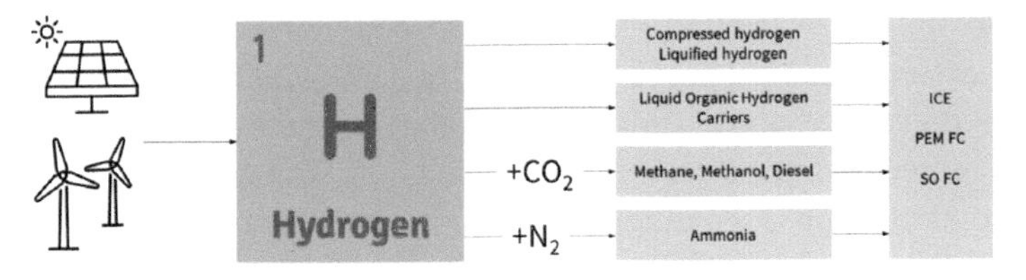

Figure 5.11 Hydrogen and hydrogen-based fuel options (Grzegorz Pawelec, 2021).

will not help reduce carbon emissions. In addition, when it comes to energy use onboard, hydrogen mixed with other marine fuels or fed to auxiliary machinery can partially reduce decarbonization in maritime transport.

Aerospace, chemical, electronics/semiconductor, and metalworking industries use liquid hydrogen (NCE, 2019). Hydrogen has a high specific energy in joules or kWh/kg but a low energy density compared to other fuels for maritime transport. By reducing the temperature of the hydrogen to – 252,9 °C, it transforms into a liquid form, which is more appropriate for the distribution of large amounts. Liquid hydrogen at 1 bar includes about four times the energy per volume unit as compressed hydrogen at 250 bars and almost three times as much as at 350 bars. Since natural gas is the primary source of hydrogen, most hydrogen liquefaction plants are located at natural gas terminals and liquefy hydrogen at the site where hydrogen is made. Also, hydrogen is best produced near demand points, as it is expensive to transport.

In the last 10 years, liquid hydrogen has been considered a fuel for ships. Hydrogen on board can be combusted directly in an internal combustion engine or used to generate electricity in fuel cells. Hydrogen combustion can result in NO_x formation, which does not occur in fuel cells. However, it should be remembered that fuel cells produce electricity directly, while internal combustion engines (ICE) primarily produce mechanical energy. So, when electricity is needed, the ICE must first turn the energy in the fuel into mechanical energy and then into electrical energy, which makes it even less efficient (Grzegorz Pawelec, 2021). Liquid hydrogen can be refilled at stations on land, but there has yet to be a solution for maritime bunkering. Liquid hydrogen bunkering on ships is not regulated, but hydrogen refueling stations have protocols and codes.

Fuel cells can transform the chemical energy of fuels into electrical and thermal energy. Maritime stakeholders endeavor safe design, construction, installation, and operation of fuel cells. As fuel cell technology has improved, the IMO has sped up its work to create ship safety procedures within the regulatory framework (IMO, 2022, 2018). Ships have high capacities for their main propulsion systems. Fuel cells are better for inland navigation and short-sea shipping (De-Troya et al., 2016; DNV, 2021b) because they require limited installed power, falling within the range currently available in fuel cells. The technology to combine fuel cells on larger ships, like cruise ships and containerships, is being developed quickly and changed to meet their needs. Fuel cells are mainly envisaged to power the auxiliary systems of larger vessels, offering a zero-emissions solution for ships idling at port or using auxiliary power for the moment. The next major technological push will entail scaling up ships' primary propulsion systems to full power (DNV, 2021b). In order to get certification for fuel cell systems, shipyards, and equipment makers must meet specific safety requirements. Once fuel cells are added to a ship, the crew must be kept safe, and the fuel cell equipment must be used correctly. For this purpose, classification societies defined fuel cell-powered ship design components and requirements (BV, 2022). Figure 5.12 shows the system components of the fuel cell on board.

The European Maritime Safety Agency (EMSA) and DNV have evaluated different fuel cell technologies. According to the evaluation of physical size, power levels (kW), relative cost, lifetime, tolerance for cycling, flexibility toward the type of fuel, technological maturity, sensitivity to fuel impurities, emissions, safety, and efficiency, the solid oxide fuel cell, the proton exchange membrane fuel cell (PEMFC), and the high-temperature PEMFC are the three most promising technologies for marine use (DNV-GL, 2010). PEMFC successfully employed in marine and high-energy demand applications stands out as an advanced technology (Shih et al., 2014). In PEMFCs, the process occurs with pure hydrogen at low operating temperatures. The primary safety factors are thus related to the service and storage of hydrogen on a vessel (Li et al., 2018; Van Hoecke et al., 2021). Susceptibility to impurities such as sulfur and CO in hydrogen, a complex water management system (both gas and liquid), and average lifetime are the main disadvantages of PEMFC technology in the energy conversion from hydrogen to electricity. Despite these disadvantages, PEMFC technology received the highest rankings by

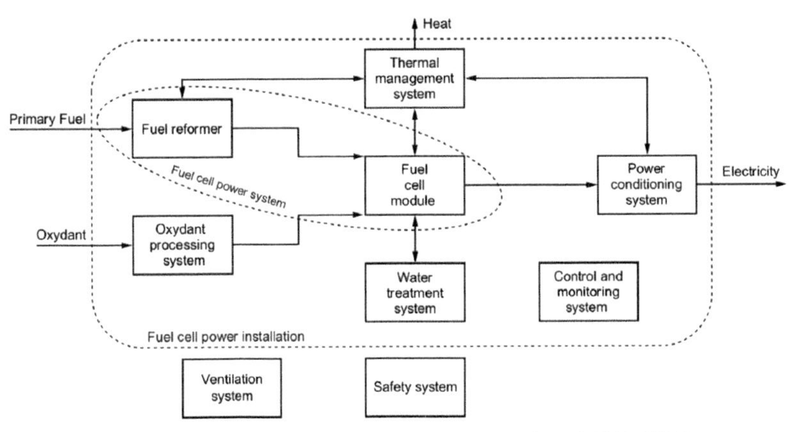

Figure 5.12 Fuel-cell system components onboard (BV, 2022).

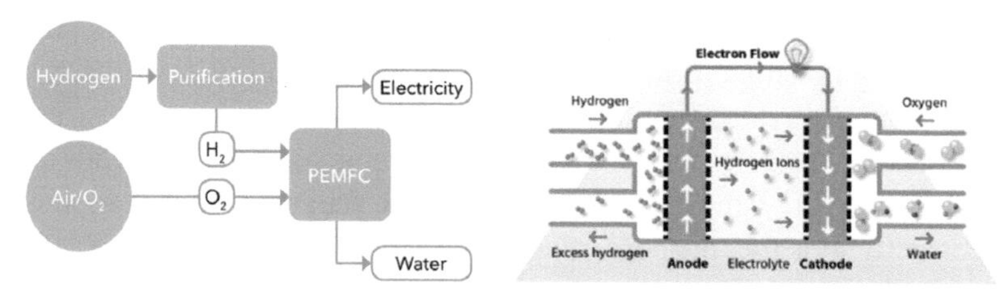

Figure 5.13 Flowchart and schematic of a PEMFC (DNV-GL, 2010).

EMSA and DNV. A flowchart and schematic for a PEMFC using hydrogen are given in Figure 5.13.

Platinum-based electrodes are used in PEMFC. The electrolyte is a moistened polymer membrane with an electrical insulator that lets hydrogen ions (H+) pass through. PEMFC uses hydrogen and oxygen to produce electricity; heat and water are obtained when the two react. The primary reactions that are emerging are:

Anode reaction: $2H_2 \rightarrow 4H^+ + 4e^-$

Cathode reaction: $O_2 + 4H^+ + 4e^- \rightarrow 4H_2O$

Total reaction: $2H_2 + O_2 \rightarrow 2H_2O$

PEMFCs are continually being developed to increase operational flexibility and durability and reduce cost. This development includes metal-organic frameworks as a new membrane material and reducing catalyst loading.

5.5 Hydrogen Bunkering from Offshore Wind Power Plants

Offshore wind power can produce "green hydrogen" or "green hydrogen-based fuels" for marine applications. Offshore wind power plants could pave the way for zero-emission ship fuels such as hydrogen, which cannot meet the energy demands of deep-sea vessels due to storage problems and are currently unavailable due to difficulty in refueling. Offshore wind power plants can be bunkering stations for ocean-going vessels globally. Figure 5.14 illustrates the offshore hydrogen bunkering system. An offshore wind power plant would use a floating structure as a bunkering station. A floating structure can accommodate a seawater purification and liquefying system, an electrolyzer, and storage and bunkering facilities to supply green hydrogen. Although hydrogen produced by offshore wind energy is more costly than conventional production methods today, efforts to reduce emissions worldwide will soon make it essential.

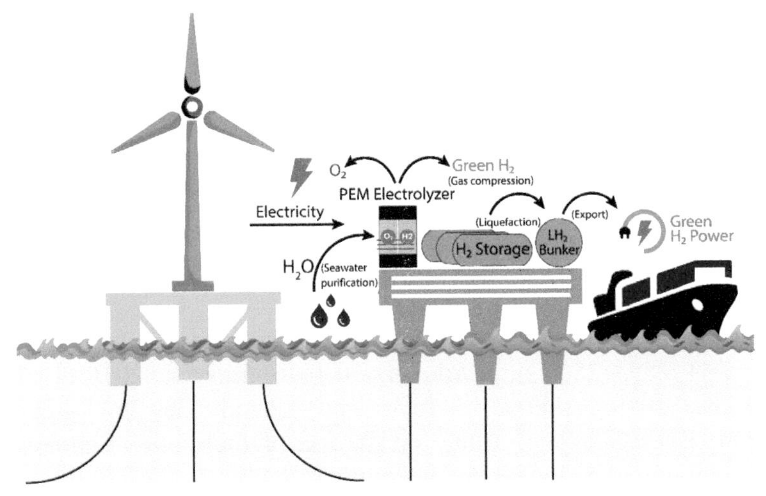

Figure 5.14 Hydrogen bunkering from the offshore wind power plant.

5.6 Conclusion

According to this chapter, many authorities anticipate that ships and their auxiliary equipment will soon run on hydrogen or hydrogen-based synthetic fuels to meet the shipping industry's zero-emission goal. Hydrogen PEMFCs are candidates for the main power supply for ships' primary propulsion and auxiliary systems, although today, their capacity is too low to power large merchant ships. However, for ships to have zero emissions, they need to use hydrogen that is made in a green way. Offshore wind energy is an essential source for the mass production of green hydrogen. It is a promising solution both for producing green hydrogen from seawater by electrolysis and having the potential to act as a refueling station for ships since hydrogen cannot be stored on board in large quantities. Thus, offshore wind power could pave the way for zero-emission long-distance maritime transportation by eliminating the need to store large amounts of hydrogen on board.

References

[1] ABS, 2020. Production of green hydrogen.

[2] British Petroleum, 2020. Energy Outlook 2020 edition explores the forces shaping the global energy transition out to 2050 and the surrounding that. BP Energy Outlook 2030, Stat. Rev. London Br. Pet. 81.

[3] BV, 2022. Ships using Fuel Cells.Caine, D., Wahyuni, W., Pizii, B., Iliffe, M., Ryan, B., Bond, L., 2021. ERM Dolphyn Hydrogen Phase 2 final report.

[4] Calado, G., Castro, R., 2021. Hydrogen production from offshore wind parks: Current situation and future perspectives. Appl. Sci. 11. https://doi.org/10.3390/app11125561

[5] Cecchinato, M., Ramírez, L., Fraile, D., 2021. A 2030 Vision for European Offshore Wind Ports.

[6] De-Troya, J.J., Álvarez, C., Fernández-Garrido, C., Carral, L., 2016. Analysing the possibilities of using fuel cells in ships. Int. J. Hydrogen Energy 41, 2853–2866. https://doi.org/10.1016/j.ijhydene.2015.11.145

[7] DNV-GL, 2019. Alternative fuels and technologies for greener shipping.DNV-GL, 2010. Study on the use of fuel cells in shipping. https://doi.org/10.5220/0002697000590063

[8] DNV, 2022. Floating Offshore Windă: The Next Five Years.DNV, 2021a. Energy Transition Outlook 2020 - A global and regional forecast to 2050, Energy Transition Outlook.

[9] DNV, 2021b. Maritime Forecast To 2050 - Energy Transition Outlook 2021 118.DNV GL-Maritime, 2019. Assessment of Selected Ternative Fuels and Technologies.

[10] ERM, 2019. Dolphyn Hydrogen Phase 1 - Final Report, Low Carbon Hydrogen Supply Programme.

[11] European Union, 2016. FOWIND - Supply chain , port infrastructure and logistic study for offshore wind farm development in Gujarat and Tamil Nadu.

[12] Fahnestock, J., Bingham, C., 2021. Mapping of Zero Emission Pilots and Demonstration Projects, Getting to Zero Coalition.

[13] GL Garrad Hassan, 2013. A guide to UK offshore wind operations and maintenance. Scottish Enterp. Crown Estate 42.

[14] Grzegorz Pawelec, 2021. Comparative report on alternative fuels for ship propulsion, Interreg North-West Europe H2 Ships.

[15] Ibrahim, O.S., Singlitico, A., Proskovics, R., McDonagh, S., Desmond, C., Murphy, J.D., 2022. Dedicated large-scale floating offshore wind to hydrogen: Assessing design variables in proposed typologies. Renew. Sustain. Energy Rev. 160, 112310. https://doi.org/10.1016/j.rser.2022.112310

[16] IMO, 2022. MEPC-78-7-2-Revision-of-the-Initial-IMO-GHG-Strategy.

[17] IMO, 2020. Fourth IMO GHG Study 2020, MEPC 75/7/15.

[18] IMO.IMO, 2018. RESOLUTION MEPC.304(72) (adopted on 13 April 2018) INITIAL IMO STRATEGY ON REDUCTION OF GHG EMISSIONS FROM SHIPS.

[19] International Energy Agency, 2020. The role of CCUS in low-carbon power systems. role CCUS low-carbon power Syst.

[20] Jackson, C., Davenne, T., Makhloufi, C., Wilkinson, I., Fothergill, K., Greenwood, S., Kezibri, N., 2020. Ammonia to green hydrogen project, Feasibility study.

[21] Jiang, Z., 2021. Installation of offshore wind turbines: A technical review. Renew. Sustain. Energy Rev. 139, 110576. https://doi.org/10.1016/j.rser.2020.110576

[22] Leanwind, 2015. Logistic Efficiencies And Naval architecture for Wind Installations with Novel Developments Description of an 8 MW reference turbine.

[23] Lee, J., Zhao, F., 2021. GWEC Global Wind Report, Global Wind Energy Council.

[24] Lerch, M., De-Prada-Gil, M., Molins, C., 2021. A metaheuristic optimization model for the inter-array layout planning of floating offshore wind farms. Int. J. Electr. Power Energy Syst. 131, 107128. https://doi.org/10.1016/j.ijepes.2021.107128

[25] Li, F., Yuan, Y., Yan, X., Malekian, R., Li, Z., 2018. A study on a numerical simulation of the leakage and diffusion of hydrogen in a fuel cell ship. Renew. Sustain. Energy Rev. 97, 177–185. https://doi.org/10.1016/j.rser.2018.08.034

[26] Musial, W., Butterfield, S., Ram, B., 2006. Energy from offshore wind. Offshore Technol. Conf. 2006 New Depths. New Horizons 3, 1888–1898. https://doi.org/10.4043/18355-ms

[27] NCE, 2019. Norwegian future value chains for liquid hydrogen. NCE Marit. CleanTech 89.

[28] Pavithran, A., Sharma, M., & Shukla, A. K. (2021). Oxy-fuel Combustion Power Cycles: A Sustainable Way to Reduce Carbon Dioxide Emission. Distributed Generation & Alternative Energy Journal, 335-362.

[29] Shih, N.C., Weng, B.J., Lee, J.Y., Hsiao, Y.C., 2014. Development of a 20 kW generic hybrid fuel cell power system for small ships and underwater vehicles. Int. J. Hydrogen Energy 39, 13894–13901. https://doi.org/10.1016/j.ijhydene.2014.01.113

[30] Smith, T., Baresic, D., Fahnestock, J., Galbraith, C., Velandia Perico, C., Rojon, I., Shaw, A., 2021. A Strategy for the Transition to Zero-Emission Shipping.

[31] Stori, V., 2021. Offshore Wind to Green Hydrogen Insights from Europe. Sumer, B.M., Kirca, V.S.O., 2022. Scour and liquefaction issues for anchors and other subsea structures in floating offshore wind farms: A review. Water Sci. Eng. 15, 3–14. https://doi.org/10.1016/j.wse.2021.11.002

[32] Van Hoecke, L., Laffineur, L., Campe, R., Perreault, P., Verbruggen, S.W., Lenaerts, S., 2021. Challenges in the use of hydrogen for maritime applications. Energy Environ. Sci. 14, 815–843. https://doi.org/10.1039/d0ee01545h

[33] Wind Energy Ireland, GreenTech Skillnet, G.& D.G.L., 2022. Hydrogen and Wind Energy.

[34] WindEurope, 2021. Ports as Key Players in the Offshore Wind Supply Chain.

[35] World Bank, 2019. Going Global-Expanding Offshore Wind to Emerging Markets. Esmap 1–44.

6

Liquid Organic Hydrogen Carrier System

Surbhi Sharma and Khushbu Gumber

Chemistry Department, University Institute of Sciences, Chandigarh University, India
E-mail: surbhi.e9299@cumail.in

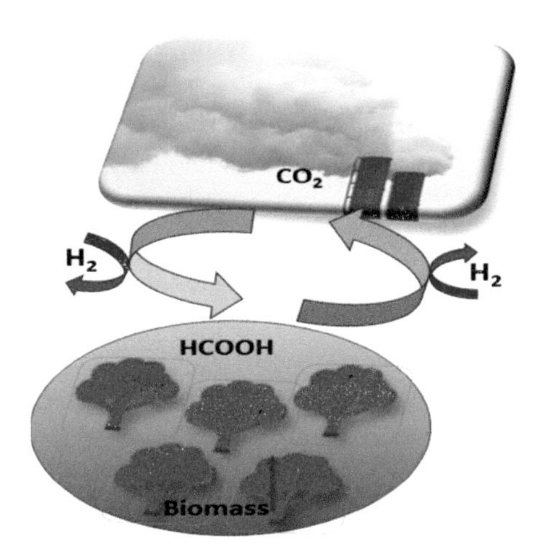

Abstract

The need for energy has increased considerably around the world because of industrialization, population growth, and the use of fossil fuels running out. So, the need for a sustainable energy supply is one of the most important socioeconomic problems the world faces in the 21st century, and any renewable energy sources that could replace fossil fuels are now essential to solving this problem. Hydrogen is the best-known organic liquid system, and it has received a lot of attention because of the current "energy crisis" and concerns about the environment. However, since hydrogen has a very low energy density per volume (10 kJ/L), it is not appropriate for the storage or transportation of massive quantities of energy. This has forced researchers to identify a favorable hydrogen carrier for the "hydrogen economy," which is not only capable of storing H_2 physically or chemically in high concentration but also capable of discharging it readily on demand. Liquid organic hydrogen carriers (LOHCs) that employ a "green hydrogen production" approach (i.e., catalytic H_2 production using an earth-abundant, nontoxic, as well as cost-effective metal-based catalyst) represent an emerging technology for a creative future energy source. Common LOHCs like ammonia and hydrazine are usually poisonous, flammable, and explosive, which restricts their practical application. Formic acid (FA), which has 4.4% by weight hydrogen, is currently gaining favor as an LOHC due to its stability, low toxicity, biodegradability, practical storage and transport, and propensity for on-demand H_2 release. Fortunately, the only by-product of FA dehydrogenation is CO_2, which is effortlessly recovered and catalytically reduced back to FA, making it feasible to frequently utilize H_2 by concluding the dehydrogenation–hydrogenation cycle. However, the last process presents a thermodynamic issue. Metal-based homogeneous or heterogeneous catalysts are the best option for such reactions; yet, heterogeneous catalytic systems offer a convenient alternative over homogenous ones by providing ease in their recovery for being reused in reactions, as the latter has inherent recyclability issues, and the use of expensive ligands and additives hinders their industrial-scale application. The decisive design of metal nanostructures, however, which enables the development of high-performance materials with tailored properties to accomplish target reactions, offers the greatest difficulties for controlling the activity and specificity of heterogeneous catalysts. Recently, heteroatom-doped graphene and carbon systems in combination with transition metals have been demonstrated to be efficacious heterogeneous catalyst systems that govern a variety of uses, including reductions, oxidations, C–C bond formation, etc.

Herein, with a rational strategy toward the mitigation of the "energy crisis," the novel methods and strategies for a hydrogen economy are summarized. The modernist perspective on the emergence of formic acid as a hydrogen generator will be emphasized in this chapter. Furthermore, its decomposition using heterogeneous and homogeneous catalysts received a good deal of attention. The chapter's conclusion offers an in-depth summary of the necessary actions that must be taken in order to lay the groundwork for a hydrogen economy based on FA.

Keywords: Liquid organic hydrogen carriers, formic acid, heterogeneous catalysts, homogeneous catalysts, cyclohexanes, N-heterocycles

6.1 Introduction

The modernization of society has made energy supplies scarce. The International Energy Agency (IEA) was created to be at the center of the global energy narrative, investigating the full range of energy options to guarantee their availability, dependability, and sustainability. In this respect, the government has created a strategy in line with its recognition of the significance of alternatives for generating power. Additionally, the scientific community is working nonstop to find ways to meet the continuously increasing global energy demand through sustainable and ecologically friendly solutions (Oliver, 2008). Traditional energy sources like fossil fuels quickly exhaust us and cause significant environmental damage, including significant greenhouse gas problems. Additionally, it is anticipated that the fossil fuel supply will not be sufficient to satisfy the total energy demand, necessitating further study of renewable energy sources (Shoko et al., 2006).

Hydrogen is being looked at as a possible power source that can be used in fuel cells to make carbon-free energy. It can be used to carry energy and store it so that it can be delivered when needed (Armaroli & Balzani, 2011).

When Bockris first thought about using hydrogen as a source of energy and expected moderation in 1972, the main problems were that people did not know much about hydrogen and there were not enough qualified engineering professors. So, it was pretty clear that one of the biggest problems with building a sustainable hydrogen economy is the lack of a safe and effective way to move, handle, and store hydrogen (Bockris, 1981). With the major storage methods like pressurization and/or cryogenic liquefaction, it is not possible to store as much as possible and do it as efficiently as possible at the same time. Because of this, the best way to store hydrogen as a liquid using molecules that can release hydrogen, such as borohydrides, hydrazine, and most recently formic acid (FA), has become the most studied area of green and clean energy (Müller et al., 2017). Also, since electricity cannot be stored in large amounts for future use, it would be safer and easier for energy projects to store chemicals.

Formic acid (FA), which is being looked into because it might be able to make hydrogen, has become one of the most interesting and important compounds in terms of storage (Schmidt et al., 2014). It has been found that FA releases hydrogen through a process that is not exothermic but rather exergonic. This thermodynamic condition causes H_2 to release from the liquid carrier at low temperatures (T = 80 °C) and high H_2 pressures (P > 600 bar), among other unique advantages. The formal energy density of FA's 53 g H_2/L, or 1.76 kWh/L, is 6.41 MJ/L.

A well-known student team initiative called "FAST" from the Netherlands created a bus using formic acid as an alternative fuel to electric or fossil fuels as part of their energy transition challenge in 2016 (www.teamFAST.nl; Huijben et al., n.d.). As a result, liquid FA presents a potential chance for long-term energy storage. Additionally, as shown in Figure 6.1, the liquid state of formic acid may offer an effective method of H_2 delivery to H_2 refueling stations.

Since transition metals have been used for a long time to lower the activation energy of reaction pathways, the catalytic strategy in many organic reactions, such as hydrogenation and dehydrogenation, has been damaged. Based on this simple idea, researchers are carefully making strong metal-based catalysts with a variety of ligands to improve the selectivity and efficiency of catalytic performance for a specific reaction. Using the idea of metal-based catalytic dehydrogenation and reversible hydrogenation (for storage) would make using FA to store hydrogen more interesting in terms of efficiency and selectivity. Due to this, ongoing reports or articles on catalysis currently support the majority of recent research on this subject (Li et al., 2017; Himeda, 2009). But most catalytic processes give a more accurate description of how hydrogen is released through catalysis, while the hard hydrogenation of carbon dioxide to FA has been studied less (Figure 6.2). Recently, there have been fewer reports published, which attempt to cover both aspects.

As far as we know, no book chapter has done a thorough comparison of different processes for LOHCs, such as the production of FA (mostly through

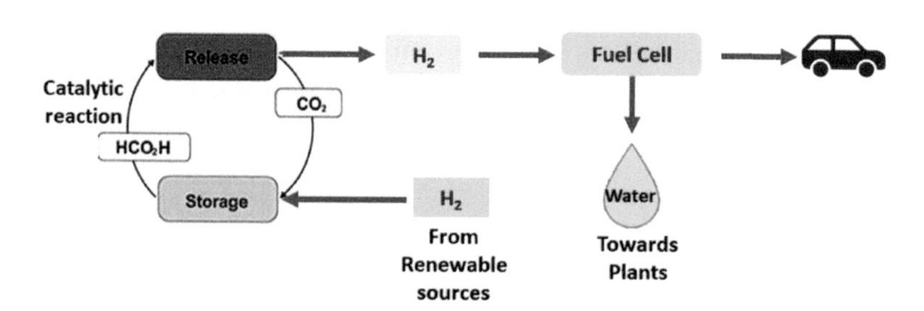

Figure 6.1 Proposed schematic diagram for formic acid-based hydrogen carrier system.

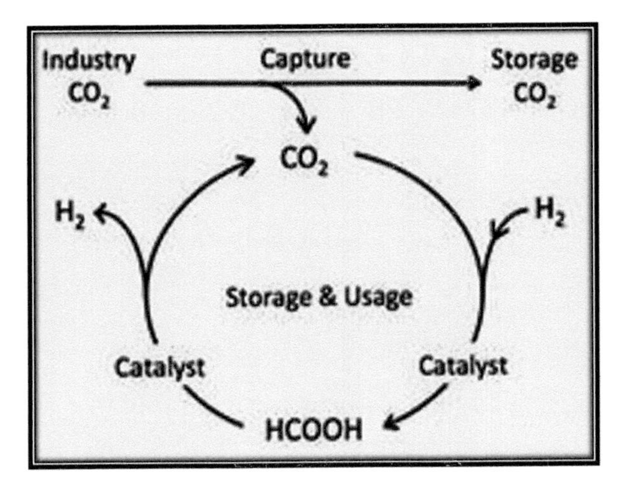

Figure 6.2 Formic acid storage strategy for H_2 production via utilizing CO_2.

catalysis), the recovery of H_2 from FA, and the purification of hydrogen as a clean source of power. The information that is summed up in this section on "liquid organic hydrogen carrier systems" is very important for figuring out if FA could be used as a reliable way to transport and store hydrogen. The present chapter could be used in designing and bridging the gap between structure-activity relations and catalytic performance, with the goal of a systematic assessment of the various chemical processes of FA-based hydrogen storage.

6.2 Shortcomings in Formic Acid-based Fuel Cells and Advantages of Catalytic Degradation of Formic Acid into H_2

The fuel cell is an extensively utilized direct chemical energy delivery system. Similar ideas apply to formic acid as well. Literature notifications have verified the efficiency and chemistry of FA as a fuel in direct-formic acid fuel cells (DFAFCs). The chemical reactions occurring in DFAFCs require two electron oxidations of FA (at the anode) and two electron reductions of O_2 (at the cathode) (eqn (6.1)–(6.3)). However, in addition to the catalyst deactivation and gasoline crossover, DFAFCs also suffer the following drawbacks: specifically, poisoning the catalysts with CO via the disadvantageous route of FA dehydration (eqn (6.4)) destroys the fuel cell at 20 ppm (Aslam et al., 2012).

Anodic reaction:

$$\text{HCOOH} \rightarrow \text{CO2} + 2\text{H} + +2\text{e} - \text{EO} \sim -0.25\ \text{V} \tag{6.1}$$

Cathodic reaction:

$$1/2\text{O2} + 2\text{H} + +2\text{e}- \rightarrow \text{H2OEO} = 1.23\ \text{V}(6.2) \tag{6.2}$$

Overall reaction:

$$\text{HCOOH} + 1/2\text{O2} \rightarrow \text{CO2} + \text{H2O EO Cell} \sim 1.48\ \text{V}(6.3) \tag{6.3}$$

Undesired route:

$$\text{HCOOH} \rightarrow \text{CO} + \text{H2O}(6.4) \tag{6.4}$$

The method also has the problem that FA can damage the active phase of the catalyst, and FA's water-loving nature can dry out the proton exchange membrane (PEM), which makes the cell more resistant. On the other hand, formic acid can be converted into hydrogen by a catalytic process under mild conditions, and the process is extremely accessible, repeatable, and useful for storage. Therefore, with the immense dedication of researchers from all over the world, notable progress has been made in recent years on the effective use of formic acid as a renewable source of energy via catalytic hydrogen release.

6.3 Decomposition of Formic Acid

Fuel cells were continuously blooming in the demand of clean energy source due to the issue of producing, storing and transporting hydrogen affordably. Rather transporting hydrogen gas, it would be more practical concept to possess a hydrogen-containing material like a chemical hydrogen storage device that can release hydrogen under ambient/simple conditions to generate H2 gas whenever required. FA, containing 4.4 wt.% hydrogen, ideally decomposed into two pathways (Eq. (1) and (5)),- CO2 and H2 (5, desired reaction) and CO and H2O (6, undesired side reaction)

$$\text{HCOOH}\rightarrow H_2 + CO_2 \Delta G = -32.9\ \text{kJmol}^{-1} \tag{6.5}$$

$$\text{HCOOH}\rightarrow \text{H}_2\text{O} + \text{CO}\Delta\text{G} = -28.5\ \text{kJmol}^{-1} \tag{6.6}$$

As the idea of using CO_2 and FA together as a hydrogen storage system may be simple and elegant, Preuster et al. (2017) say that the CO-free decomposition of formic acid is very important for the design and operation of FA-based hydrogen storage systems. This strategy could be explained by the fact that formic acid breaks down selectively into H_2 and CO_2 and that CO_2 is reused when the amount of H_2 in formic acid goes down. Furthermore, the earth's abundant, low-cost, and simple CO_2 supply favors this chemical reaction. Highly motivated research teams are working at all times to develop the most appropriate homogeneous or heterogeneous catalysts for the selective decomposition of FA as a clean energy weapon (Savinell, 1998; Kariya et al., 2003). To fully achieve selectivity for the decomposition of formic acid along the desired pathway, however, is still an uphill battle.

6.4 Homogeneous Catalyst for Selective HCOOH Decomposition

Researchers are working around the clock to find the best catalysts (either homogeneous or heterogeneous) for the selective breakdown of FA as a weapon for clean energy. The desired pathway still needs to be found, which is hard. Only then will the breakdown of formic acid be completely selective. One of the most recent reports on organometallic catalysts presented soluble Pt, Ru, and Ir phosphine complexes for selective FA decomposition and found that the iridium complex $IrH_2Cl(PPh_3)_3$ had the highest rate of

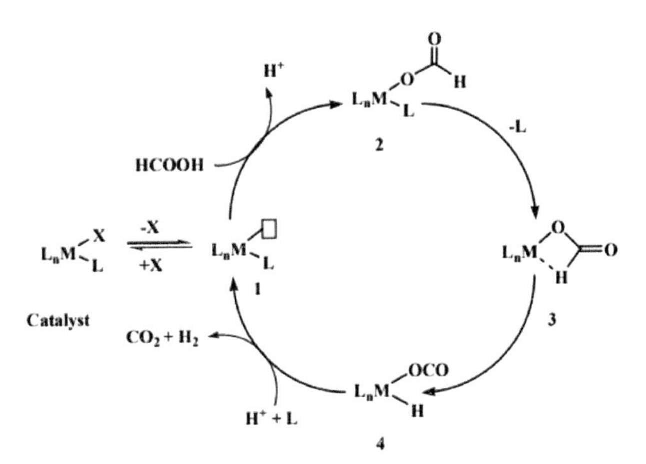

Figure 6.3 General proposed mechanism of formic acid decomposition.

decomposition among all the complexes. $Rh(C_6H_4PPh_2)(PPh_3)_2$ is active for the decomposition of formic acid (Coffey, 1967; Savinell, 1998; Kariya et al., 2003). According to reports on organometallic catalysts, Rh and Ru metals have been extensively utilized to create homogeneous catalysts that decompose FA efficiently (Fellay et al., 2008; Boddien et al., 2009). Researchers working on the hydrogen economy have also investigated the catalytic activity of Rh using a variety of compounds, including $[RhCl_2(p\text{-}cymene)]_2$, $[RhCl_2(PPh_3)_3]$, $[RhCl_3(benzene)_2]$, and $RhBr3_xH_2O$, in the presence of various ligands and amines. According to Junge et al. (2009), both bromide and chloride ion-based precursors exhibit increased catalytic activities, with TOFs greater than 300 h^{-1} at 40 °C and no evidence of CO as a by-product (Loges et al., 2008).

Later, it was said that a $RuCl_2(benzene)_2$-based soluble catalyst for FA breakdown had a turnover number (TON) of 2,60,000 and an average TOF of 900 h^1 (Boddien et al., 2009). Additionally, one report of accelerated catalytic activity in the presence of $Ru(H_2O)_6$ or $RuCl3_xH_2O$ customized triphenylphosphine derived ligand was presented. This was achieved by incorporating sodium formate into the aqueous FA solution. The most advantageous aspect of the method is that, even after more than 100 cycles at 90 °C, neither the catalyst nor CO were detected in this system. After these successful stories behind noble metal-based catalysts, the high price of noble metals motivated the scientists to seek alternate environmentally benign, selective, and economically favorable transition metal-based catalysts for formic acid decomposition (Figure 6.3). Later, a number of efficient Fe-based homogeneous catalysts for FA decomposition were developed, with a pincer-supported (pincer ligand can provide support and firm structure to the catalyst) Fe catalyst exhibiting a high TON being reported (Bielinski et al., 2014).

6.5 Heterogeneous Catalyst for Selective HCOOH Decomposition

Even though there are enough examples of homogenous catalysis working, researchers are now focusing on making heterogeneous catalysts or making homogenous catalysts more heterogeneous to make them more active and specific. Au, Ir, Pd, Pt, Rh, and Ru, among others, were mostly investigated in recent studies as noble monoatomic metal catalysts for the catalytic decomposition of formic acid. Nanosized Au particles were created over Al_2O_3 by Ojeda and Iglesia, who also witnessed a highly active decomposition reaction without CO evolution (Ojeda & Iglesia, 2009). A gas phase synthesis of commercial Pd/C, Au/TiO_2, and Au/C was given in a different study (Bulushev et al., 2010). Additionally, it was discovered that substituting Pd for Au or Ag could greatly increase the catalyst's activity and selectivity while also changing the active Pd alloy's electronic properties (Zhou et al., 2008). The most recent study on a novel metal organic framework-based catalyst for FA decomposition introduced the first extremely active Au-Pd embedded MOF (Gu et al., 2011).

Furthermore, catalysts with various structures showed exceptional qualities along with metals. Researchers recently reported on this, synthesizing PdAu@Au core shell nanomaterials over activated carbon with significantly increased activity and durability compared to the Au/C and Pd/C nanocomposites (Wang et al., 2013). In a different research, noble metals were investigated in order to create catalysts for the breakdown of FA that had multiple components and were highly effective. To improve efficiency, fine nano-Pd particles were immobilized on diamine-alkali-reduced graphene oxide (Pd/PDA-rGO). These composites were discovered to have the greatest TOF value for the breakdown of aqueous formic acid (3810 h^{-1} at 50 °C). The inventory of additional heterogeneous catalysts for FA decomposition, particularly in the aqueous phase, is provided in Table 6.1.

6.6 Alternative LOHC Systems

In many ways, hydrogen generation reactions have been utilized. Early studies on these liquid organic hydrogen storage materials (Hodoshima et al., 2003) concentrated on using various cycloalkanes, such as cyclohexane, methylcyclohexane, and decalin, among others. This dehydrogenation of

Table 6.1 Preferred heterogeneous catalysts for the decomposition of formic acid.

Catalyst	Reaction condition/temperature	TOF (h^{-1})
Pd/C	25	64
AgPd/C	50	382
AuPd/C	50	230
AuPd/ED-MIL-101	90	106
$Co_{0.3}Au_{0.35}Pd_{0.35}/C$	RT	80
PdNi@Pd/GNs-CB	RT	150
CoAuPd/DNA-rGO	RT	85
Pd/PDA-rGO	50	3810

cycloalkanes to their corresponding aromatics only has one disadvantage: it requires comparatively high temperatures due to the unfavorable enthalpy changes that result. Another literature study (Okada et al., 2006) showed that the introduction of heteroatoms (N or B) in a cyclic system reduces the energy input in dehydrogenation, indirectly favoring the enthalpy change. This fact is further supported and validated theoretically by Biniwale et al. (2008) who enclosed a direction for remarkable investigations on heterocycles (Teichmann et al., 2012). Some of the hydrogen storage systems other than formic acid are discussed below in detail.

6.6.1 Cycloalkanes

Cycloalkanes have been looked at for a long time as a source of liquid hydrogen because they are nontoxic, can be liquid, have a high boiling point, produce more hydrogen (6%–8% by weight), and, most importantly, do not produce CO_2 or CO. Because the process of dehydrogenating cycloalkanes to get the hydrogen out is very endothermic (63–69 kJ/mol H_2), cycloalkanes are still not good for transporting or storing liquid hydrogen (Shukla et al., 2010; Wang et al., 2008). The additional analysis revealed that this disadvantage can be somewhat remedied by conducting a systematic study (Figure 6.4) on the substituted cycloalkanes, as it was discovered that methylcyclohexane dehydrates more readily than cyclohexane due to the electron-donor methyl substituent (Okada et al., 2006).

Based on a different research, decalin and methylcyclohexane have different thermodynamically advantageous dehydrogenation energies. The literature (Yolcular & Olgun, 2008; Riad & Mikhail, 2008) provides detailed explanations of a variety of catalytic dehydrogenation processes employing noble metals like Pt for dehydrogenation of cycloalkanes (shown in Table 6.1). The only drawback of the process is that coking makes it challenging to deactivate extremely active catalyst. According to Okada et al. (2006) and Lázaro et al. (2008), the introduction of the additional metal (Ir, Re, Rh, Pd, W, etc.) combined with a promoter (Ca) or the introduction of a suitable support material (CNF and Al_2O_3) is successful against coking. Other than these noble metals, some other metals like Ni- and Mo-based catalysts are also explored for dehydrogenation owing to their lesser costs and similar activities(Yolcular & Olgun, 2008; Riad & Mikhail, 2008).

The sextet mechanism and the doublet mechanism are the two main theories that underpin the various proposed mechanisms for the catalytic dehydrogenation of cyclohexane and related molecules. In the first scenario, cyclohexane is adsorbed on the catalyst surface and proceeds through direct dehydrogenation into the benzene ring system, whereas in the second scenario, hydrogen is released progressively because of side adsorption through the C=C double bond on the catalyst surface (Figure 6.5).

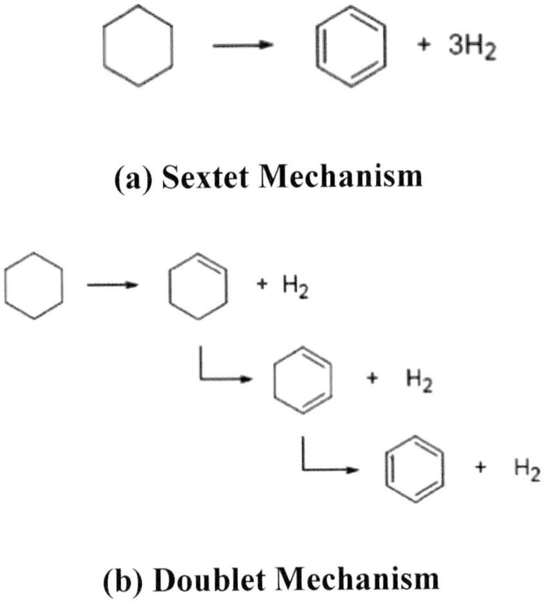

+ $3H_2$

(a) Sextet Mechanism

+ H_2

+ H_2

+ H_2

(b) Doublet Mechanism

Figure 6.4 Proposed reaction mechanism for the dehydrogenation of cyclohexane.

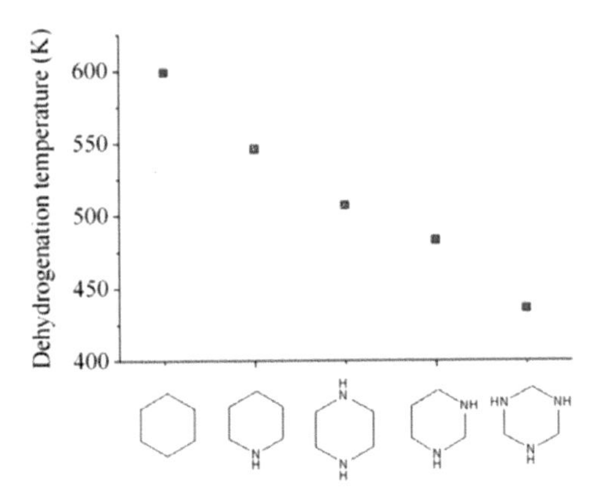

Figure 6.5 Dependency of dehydrogenation temperature of the cyclohexane on N content.

Extremely active catalysts are necessary to facilitate the hydrogenation and dehydrogenation of LOHCs. Two model compounds, toluene and benzene, were examined for hydrogenation in order to comprehend the performance of the catalysts. Recently scientists reported a new slurry form of catalysts that include different metal alloys ($LaNi_5$, $MlNi_5$, Mg_2Ni, and Raney-Ni) and an unsaturated liquid hydrogen carrier to catalyze the process of hydrogenation with an add-on tendency of these catalysts to absorb hydrogen by themselves under a simple reaction condition with storage of hydrogen in both metal alloys and the organic hydride (Chen et al., 2003). Raney-Ni emerged as the best alloy among the different tested alloy systems. Overall, this slurry system showed better hydrogenation performance; however, the process of dehydrogenation performance did not show any significant results as of yet. The degree of cyclohexane dehydrogenation in the $MlNi_5$-C_6H1_2 slurry system was less than 3%, which corroborated these findings (Yolcular & Olgun, 2008). Despite getting a lot of attention as a hydrogen producer and storage device, cycloalkanes are a less desirable option due to the unfavorable thermodynamic temperature requirement of over 300 °C. However, due to their low costs, high purity, and effective utilization of the bi-product benzene and toluene obtained in 5×10^7 ton/year and 1×10^7 ton/year, respectively, these cycloalkanes are ideal for the large-scale and long-distance hydrogen delivery.

6.6.2 N-Heterocycles

When heteroatoms are added to cycloalkanes, the molecule's structure changes. This makes the thermodynamic properties of the molecule much better than those of alkane-arene pairs. According to several patents (Pez et al., 2008), the first heteroatomic molecule in this area has been developed (Okada et al., 2006). They showed that a system with various conjugated heteroatoms (N, O, P, B, etc.) in a cyclic molecule can be hydrogenated or de-hydrogenated more effectively. LOHCs, along with nitrogen incorporation, that experienced favorable enthalpy changes during dehydrogenation are described in depth. In a different study, the thermodynamic analysis of several N-heterocycles revealed that the hydrogen release temperature for N-incorporated rings, particularly at positions 1 and 3, was significantly decreased. According to further investigation, N-substituted five-membered heterocycles dehydrate with lower dehydrogenation enthalpies than six-membered ring systems. By altering the enthalpy of dehydrogenation, the substitution of C with other heteroatoms like O, B, and P may also have an impact on the efficiency of LOHCs (Biniwale et al., 2008; Cui et al., 2008).

6.7 Conclusion and Outlook

Production, storage, and use of the power source must all be addressed concurrently to satisfy the current demand for hydrogen-powered transportation. Leveraging liquid organic hydrogen carriers for the effective production, storage, and transportation of hydrogen as a source of energy is one of the best options discovered to date. These sources can be thoroughly investigated regarding the reversible binding and releasing of hydrogen under various environmental circumstances using a strong metal catalyst. Formic acid, which is used to release and bind hydrogen on demand with the aid of metal-based catalysts, appears to be the most effective option among the different reported LOHCs to date. Consequently, formic acid is currently a well-researched option in the LOHC categories. Although LOHCs are rapidly becoming the most effective technology for producing and storing hydrogen, significant effort is still required to advance their implementation. The book chapter that is currently being discussed systematically discusses recent developments in various LOHCs with a primary emphasis on FA. However, additional structural studies on the design of the catalyst are required for better efficiency, cost-effectiveness, and quick reversibility in order to validate practical applications.

References

[1] Armaroli, N., & Balzani, V. (2011). The hydrogen issue. ChemSusChem, 4(1), 21–36. https://doi.org/10.1002/cssc.201000182
[2] Aslam, N. M., Masdar, M. S., Kamarudin, S. K., & Daud, W. R. W. (2012). Overview on Direct Formic Acid Fuel Cells (DFAFCs) as an Energy Sources. APCBEE Procedia, 3, 33–39. https://doi.org/10.1016/j.apcbee.2012.06.042
[3] Bielinski, E. A., Lagaditis, P. O., Zhang, Y., Mercado, B. Q., Würtele, C., Bernskoetter, W. H., Hazari, N., & Schneider, S. (2014). Lewis acid-assisted formic acid dehydrogenation using a pincer-supported iron catalyst. Journal of the American Chemical Society, 136(29), 10234–10237. https://doi.org/10.1021/ja505241x
[4] Biniwale, R. B., Rayalu, S., Devotta, S., & Ichikawa, M. (2008). Chemical hydrides: A solution to high capacity hydrogen storage and supply. International Journal of Hydrogen Energy, 33(1), 360–365. https://doi.org/10.1016/j.ijhydene.2007.07.028

[5] Bockris, J. O. (1981). A Hydrogen Economy. Comprehensive Treatise of Electrochemistry, 505–526. https://doi.org/10.1007/978-1-4615-6687-8_16

[6] Boddien, A., Loges, B., Junge, H., Gärtner, F., Noyes, J. R., & Beller, M. (2009). Continuous hydrogen generation from formic acid: Highly active and stable ruthenium catalysts. Advanced Synthesis and Catalysis, 351(14–15), 2517–2520. https://doi.org/10.1002/adsc.200900431

[7] Bulushev, D. A., Beloshapkin, S., & Ross, J. R. H. (2010). Hydrogen from formic acid decomposition over Pd and Au catalysts. Catalysis Today, 154(1–2), 7–12. https://doi.org/10.1016/j.cattod.2010.03.050

[8] Chen, C., Cai, G., Chen, Y., An, Y., Xu, G., & Wang, Q. (2003). Hydrogen absorption properties of the slurry system composed of liquid C6H6 and F-treated Mg2Ni. Journal of Alloys and Compounds, 350(1–2), 275–279. https://doi.org/10.1016/S0925-8388(02)00967-2

[9] Coffey, R. S. (1967). The decomposition of formic acid catalysed by soluble metal complexes. Chemical Communications (London), 18, 923–924. https://doi.org/10.1039/C1967000923b

[10] Cui, Y., Kwok, S., Bucholtz, A., Davis, B., Whitney, R. A., & Jessop, P. G. (2008). The effect of substitution on the utility of piperidines and octahydroindoles for reversible hydrogen storage. New Journal of Chemistry, 32(6), 1027–1037. https://doi.org/10.1039/b718209k

[11] Fellay, C., Dyson, P. J., & Laurenczy, G. (2008). A Viable Hydrogen-Storage System Based On Selective Formic Acid Decomposition with a Ruthenium Catalyst. Angewandte Chemie, 120(21), 4030–4032. https://doi.org/10.1002/ange.200800320

[12] Gu, X., Lu, Z. H., Jiang, H. L., Akita, T., & Xu, Q. (2011). Synergistic catalysis of metal-organic framework-immobilized au-pd nanoparticles in dehydrogenation of formic acid for chemical hydrogen storage. Journal of the American Chemical Society, 133(31), 11822–11825. https://doi.org/10.1021/ja200122f

[13] Himeda, Y. (2009). Highly efficient hydrogen evolution by decomposition of formic acid using an iridium catalyst with 4,4/-dihydroxy-2,2/-bipyridine. Green Chemistry, 11(12), 2018–2022. https://doi.org/10.1039/b914442k

[14] Hodoshima, S., Arai, H., Takaiwa, S., & Saito, Y. (2003). Catalytic decalin dehydrogenation/naphthalene hydrogenation pair as a hydrogen source for fuel-cell vehicle. International Journal of Hydrogen Energy, 28(11), 1255–1262. https://doi.org/10.1016/S0360-3199(02)00250-1

[15] Huijben, J. C. C. M., Wieczorek, A. J., Van Den Beemt, A. A. J., Verbong, G. P. J., & Van Marion, M. H. (n.d.). Expedition energy transitionă: lessons learned from an educational co-creation experiment Citation for published version (APA). www.tue.nl/taverne

[16] Junge, H., Boddien, A., Capitta, F., Loges, B., Noyes, J. R., Gladiali, S., & Beller, M. (2009). Improved hydrogen generation from formic acid. Tetrahedron Letters, 50(14), 1603–1606. https://doi.org/10.1016/j.tetlet.2009.01.101

[17] Kariya, N., Fukuoka, A., & Ichikawa, M. (2003). Zero-CO2 emission and low-crossover "rechargeable" PEM fuel cells using cyclohexane as an organic hydrogen reservoir. Chemical Communications, 3(6), 690–691. https://doi.org/10.1039/b211685e

[18] Lázaro, M. P., García-Bordejé, E., Sebastián, D., Lázaro, M. J., & Moliner, R. (2008). In situ hydrogen generation from cycloalkanes using a Pt/CNF catalyst. Catalysis Today, 138(3–4), 203–209. https://doi.org/10.1016/j.cattod.2008.05.011

[19] Li, Z., Yang, X., Tsumori, N., Liu, Z., Himeda, Y., Autrey, T., & Xu, Q. (2017). Tandem Nitrogen Functionalization of Porous Carbon: Toward Immobilizing Highly Active Palladium Nanoclusters for Dehydrogenation of Formic Acid. ACS Catalysis, 7(4), 2720–2724. https://doi.org/10.1021/acscatal.7b00053

[20] Loges, B., Boddien, A., Junge, H., & Beller, M. (2008). Controlled generation of hydrogen from formic acid amine adducts at room temperature and application in H2/O2 fuel cells. Angewandte Chemie - International Edition, 47(21), 3962–3965. https://doi.org/10.1002/anie.200705972

[21] Müller, K., Brooks, K., & Autrey, T. (2017). Hydrogen Storage in Formic Acid: A Comparison of Process Options. Energy and Fuels, 31(11), 12603–12611. https://doi.org/10.1021/acs.energyfuels.7b02997

[22] Ojeda, M., & Iglesia, E. (2009). Formic acid dehydrogenation on au-based catalysts at near-ambient temperatures. Angewandte Chemie - International Edition, 48(26), 4800–4803. https://doi.org/10.1002/anie.200805723

[23] Okada, Y., Sasaki, E., Watanabe, E., Hyodo, S., & Nishijima, H. (2006). Development of dehydrogenation catalyst for hydrogen generation in organic chemical hydride method. International Journal of Hydrogen Energy, 31(10), 1348–1356. https://doi.org/10.1016/j.ijhydene.2005.11.014

[24] Oliver, T. (2008). Clean fossil-fueled power generation. Energy Policy, 36(12), 4310–4316. https://doi.org/10.1016/j.enpol.2008.09.062

[25] Pez, G. P., Scott, A. R., Cooper, A. C., Cheng, H., Wilhelm, F. C., & Abdourazak, A. H. (2008). Hydrogen storage by reversible hydrogenation of pi-conjugated substrates. In U.S.

[26] Preuster, P., Papp, C., & Wasserscheid, P. (2017). Liquid organic hydrogen carriers (LOHCs): Toward a hydrogen-free hydrogen economy. Accounts of Chemical Research, 50(1), 74–85. https://doi.org/10.1021/acs.accounts.6b00474

[27] Riad, M., & Mikhail, S. (2008). Dehydrogenation of cyclohexane over molybdenum/mixed oxide catalysts. Catalysis Communications, 9(6), 1398–1403. https://doi.org/10.1016/j.catcom.2007.12.001

[28] Savinell, R. F. (1998). High Temperature Polymer Electrolyte Fuel Cells. ECS Proceedings Volumes, 1998–27(1), 81–90. https://doi.org/10.1149/199827.0081pv

[29] Schmidt, I., Müller, K., & Arlt, W. (2014). Evaluation of formic-acid-based hydrogen storage technologies. Energy and Fuels, 28(10), 6540–6544. https://doi.org/10.1021/ef501802r

[30] Shoko, E., McLellan, B., Dicks, A. L., & da Costa, J. C. D. (2006). Hydrogen from coal: Production and utilisation technologies. International Journal of Coal Geology, 65(3–4), 213–222. https://doi.org/10.1016/j.coal.2005.05.004

[31] Shukla, A. A., Gosavi, P. V., Pande, J. V., Kumar, V. P., Chary, K. V. R., & Biniwale, R. B. (2010). Efficient hydrogen supply through catalytic dehydrogenation of methylcyclohexane over Pt/metal oxide catalysts. International Journal of Hydrogen Energy, 35(9), 4020–4026. https://doi.org/10.1016/j.ijhydene.2010.02.014

[32] Teichmann, D., Stark, K., Müller, K., Zöttl, G., Wasserscheid, P., & Arlt, W. (2012). Energy storage in residential and commercial buildings via Liquid Organic Hydrogen Carriers (LOHC). Energy and Environmental Science, 5(10), 9044–9054. https://doi.org/10.1039/c2ee22070a

[33] Wang, B., Goodman, D. W., & Froment, G. F. (2008). Kinetic modeling of pure hydrogen production from decalin. Journal of Catalysis, 253(2), 229–238. https://doi.org/10.1016/j.jcat.2007.11.012

[34] Wang, Z.-L., Yan, J.-M., Ping, Y., Wang, H.-L., Zheng, W.-T., & Jiang, Q. (2013). An Efficient CoAuPd/C Catalyst for Hydrogen Generation from Formic Acid at Room Temperature. Angewandte Chemie, 125(16), 4502–4505. https://doi.org/10.1002/ange.201301009

[35] Yolcular, S., & Olgun, Ö. (2008). Ni/Al2O3 catalysts and their activity in dehydrogenation of methylcyclohexane for hydrogen production. Catalysis Today, 138(3–4), 198–202. https://doi.org/10.1016/j.cattod.2008.07.020

[36] Zhou, X., Huang, Y., Xing, W., Liu, C., Liao, J., & Lu, T. (2008). High-quality hydrogen from the catalyzed decomposition of formic acid by Pd-Au/C and Pd-Ag/C. Chemical Communications, 30, 3540–3542. https://doi.org/10.1039/b803661f

7

Hydrogen-added Natural Gas for Gasoline Engines

Nishit Bedi

Government Engineering College, Ujjain, India
E-mail: bedin732@gmail.com

Abstract

CNG has an established record of approximately three decades in the transportation sector. However, there is a need to explore a means to improve the performance of CNG-fueled engines, especially under lean burn conditions. In this context, adding small quantities of hydrogen to natural gas is observed to be effective. The behavior of hydrogen-added natural gas as a fuel for gasoline combustion engines is presented in this chapter. The performance and emission characteristics achieved for a 10% and 15% HCNG mixture are discussed. Under similar conditions, an engine was made to run on pure gasoline and neat CNG in order to compare the characteristics of the engine obtained under HCNG operation. An improvement in performance over neat CNG was observed. However, less work was produced with HCNG mixtures in comparison to gasoline. On the other hand, emission characteristics obtained for HCNG mixtures were observed to be better in comparison to pure gasoline as well as neat CNG operation.

Keywords: HCNG, hydrogen, engine, performance, emission

7.1 Introduction

The current combustion-technology-based automobiles have entered deeply into our daily lives. Presented Such vehicles are currently dealing with problems such as escalated fuel price and their adverse effect on the environment.

However, they are required to continue their key role due to their higher power-to-weight ratio in comparison to other available options such as battery-powered or fuel cell-operated vehicles (Das et al., 2000). The use of fuels other than conventional gasoline or diesel in the existing engine technology in order to run the engine in a clean manner has revealed its potential as an effective solution. Several options like alcohol, biodiesel, biogas, biomass, dimethyl ether, natural gas, and hydrogen have been discovered as fuels for future application in IC engines (Das et al., 2000; Mehra et al., 2017).

CNG has a track record of producing low-emissions for nearly three decades. Its gaseous behavior helps in resolving the problems associated with liquid fuels, such as vapor lock, cold wall quenching, inadequate vaporization, and poor mixing with air. However, a slower rate of burning and poor lean burn capability produce large cycle-to-cycle variations, resulting in lower power output and higher fuel consumption. There is a need to explore a means to enhance the flame velocity in order to improve the performance of CNG-fueled engines, especially under lean burn conditions. A small amount of hydrogen addition in CNG is an effective method to improve performance. The potential of pure hydrogen as an alternative green fuel for vehicular applications has been well investigated in the past (Colin, 2001; Eichlseder et al., 2003). Hydrogen gas can be produced from water vapor with the use of renewable energy sources. Conversely, when consumed as a fuel in the combustion process, it again generates water vapor. The acceptance of hydrogen has been observed to be much more appropriate for an SI engine rather than a CI engine because of its high auto-ignition temperature of 858 K (Das, 1996; Kahraman et al., 2007; Wang et al., 2011). Thus, we will limit our discussion to the fuel appropriate for use in gasoline engines.

7.2 Enhancement in Relevant Properties of CNG with the Addition of Hydrogen

Spark-ignition engines can be operated on a variety of fuels. Several options, such as hydrogen, gasoline, methane, and their mixtures in various percentages, can be directly used in engine systems.

The market for gasoline and methane have already been developed on a commercial basis. Hydrogen can be introduced as an additive to any of the fuels (gasoline or methane) in order to develop its market. A blend of hydrogen in natural gas seemed to be more attractive than gasoline–hydrogen blends (Kamil and Rahman, 2015).

Table 7.1 Properties of hydrogen, methane, and gasoline (White et al., 2006; Saravanan et al., 2007; Eichlseder and Klell, 2010).

Properties	Unit	Hydrogen	Methane	Gasoline
Chemical formula		H_2	CH_4	C_nH_{2n}
Molecular weight	amu	2.016	16.04	107
Density (at 1 bar, 25 °C)	Kg/m^3	0.082	0.65	750
Diffusivity in air (at 1 bar, 25 °C)	(cm^2/s)	0.61	0.16	0.005
Energy density	MJ/kg	120	50	44.8
	MJ/m^3	10.046	32.573	195.8
Stoichiometric air requirement	Kg/kg of fuel	34.3	17.2	14.7
Flammability limits (in air at 1 bar, 25 °C)	vol %	4–75	4.3–15	1.4–7.6
Flame velocity	m/s	2.65–3.25	0.37–0.45	0.37–0.43
Quenching gap in air	cm	0.064	0.21	0.2
Minimum ignition energy	mJ	0.02	0.29	0.24
Auto-ignition temperature (in air at 1 bar)	°C	585	540	247–280

Hydrogen and methane are both gaseous in behavior. They can be mixed up in any proportion. Ideal gas laws can be used to assess the properties of the prepared mixtures. The characteristics of hydrogen and methane that are related to combustion are shown in Table 7.1 (White et al., 2006; Saravanan et al., 2007; Eichlseder and Klell, 2010). The properties of gasoline are also presented for comparison.

7.2.1 Improvement in ignition energy and quenching gap

The energy required for ignition of hydrogen is observed to be approximately 15 times lower than the energy required for a methane–air mixture. Similarly, the quenching gap required during pure hydrogen operation is much lower than the corresponding gap required for natural gas or gasoline. As a result, adding hydrogen to CNG will reduce the energy requirements in order to initiate and sustain combustion.

7.2.2 Better flame velocity and flammability limit

It can be observed from the values that the rate of flame propagation of hydrogen is nearly seven times greater than that of methane in the stoichiometric fuel–air mixtures. Moreover, engines can be operated at higher excess air ratios with hydrogen due to its broader range of flammability. On the other hand, compressed natural gas is characterized by a lower flame velocity and very narrow flammability limits. The properties of methane can thus be modified by mixing hydrogen in a suitable percentage with it so as to achieve the expected burning velocity and lean burn ability of the mixture.

7.2.3 Higher flame temperature

When the engines are run on pure hydrogen, as shown in Table 7.1, the flame temperatures are higher. Such temperatures result in increased NOx emissions at the exhaust. However, the engine system can be operated with excess air due to wider limits of flammability in order to control the amount of NOx emissions.

In fact, the higher auto-ignition temperature of the hydrogen makes it extremely hard for the diesel engine to operate on hydrogen. On the other hand, the same property can be used to make a gasoline engine's compression ratio higher. Higher compression ratios help to achieve higher thermal efficiency.

The ignition energy, flame speed, limits of flammability, and flame temperature of the new fuel formed by adding hydrogen to methane will be significant in determination of performance of the system. Consequently. So, geometric parameters like the timing of the spark and the timing of the inlet and exhaust valves should be optimized based on the percentage of hydrogen added. These parameters will be quite different from the settings available for engines running on pure methane–air mixtures.

7.3 Hydrogen -CNG Blend as a Fuel for Spark-Ignition Engine

Gaseous fuels are preferred for use in combustion engines due to three main reasons. First, a homogeneous mixture is prepared that burns better than liquid fuels. Second, the smallest amount of carbon deposit can be observed upon the combustion of gaseous fuels. Third, using hydrogen as a supplement will reduce the carbon/hydrogen content of the fuel. Hence, carbon-based emissions such as CO, CO_2, and HC emissions can be minimized.

Thipse et al. (2009) suggested making some small changes to the engine system so that it could run on gaseous fuel. Consequently, a considerable increase in studies of HCNG-fueled combustion engines was observed. The characteristics of single-cylinder engines using various compositions of HCNG were investigated by Hora and Agarwal (2015). An improvement in the engine's performance characteristics with the addition of hydrogen was reported. The fuel composition of 30% hydrogen in CNG was observed to be utilized properly without any modifications to the hardware systems of the existing gas vehicles. The effect of a hydrogen–methane mixture up to 30% by volume in a four-cylinder SI engine was experimentally investigated by Akansu et al. (2007). The study was conducted for constant speed at 2000 rpm with an equivalence ratio ranging from 0.6 from 1.2. An increase in brake thermal efficiency with the addition of hydrogen to natural gas was noticed. Similar results of enhanced engine efficiency were observed in the study conducted by Das in 1996. The increase in efficiency was attributed to the improvement in combustion stability achieved due to the enrichment of hydrogen. The percentage of hydrogen up to 30% by volume in CNG was observed to demonstrate reductions in hydrocarbon emissions along with gain in thermal efficiency. However, a rise in nitrogen oxide emissions was noticed (Renny and Janardan 2008). The performance and emission characteristics of a modified Isuzu multicylinder SI engine fueled with HCNG blend were experimentally investigated by Tangoz et al. (2017). Experiments were conducted using various H2–CNG blends with hydrogen percentages ranging from 0% to 20%. Smaller values of HC and CO emissions, along with a higher amount of NOx emissions relative to general values as per Euro-5 standards, were obtained. The endurance tests conducted on a heavy-duty engine using an 18-HCNG mixture were studied by Subramanian in 2014. A decrease in the amount of CO and HC by about 40% and 25%, respectively, in comparison to a pure natural gas operation was noticed. However, an increase of about 32% in the amount of NOx emissions was also observed. Similar results of higher NOx emissions for engines operating on HCNG mixtures were also reported in other studies (Moreno et al., 2012; Tinaut et al., 2011). However, the engine can run on lean mixtures in order to control the amount of NOx and achieve higher values of thermal efficiency. Good descriptive research work in the context of HCNG blends for vehicular applications can be observed in Klell et al. (2012) and Laget et al., (2012). A state-of-the-art description of the utilization of HCNG mixtures as a fuel for combustion engines is presented in Mehra et al. (2017) and Alrazen and Ahmad (2018).

The benefit of adding hydrogen to compressed natural gas, especially in the present situation, is quite evident from the above discussion. The limits of practical interest for adding hydrogen to natural gas are shown in Figure 7.1.

It is recommended to use less than 30% of hydrogen by volume in order to avoid possible tendency of engine knock during the combustion process (Acιkgoz et al., 2015; Das and Lather, 2019). Further reduction in this

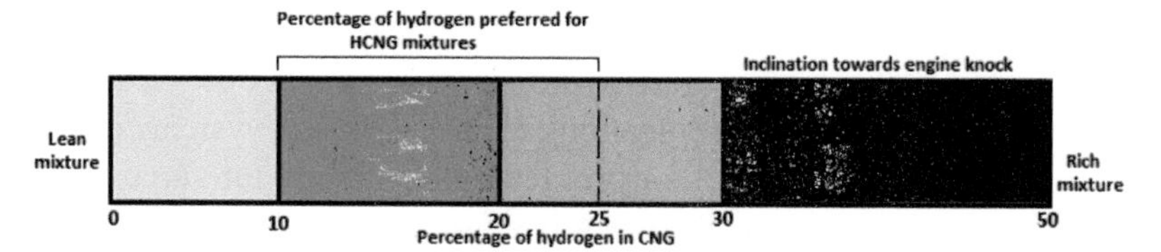

Figure 7.1 Percentage of hydrogen in CNG for its use in spark-ignition engine (Das and Lather, 2019; Mehra et al., 2017).

amount of hydrogen by about 12% was suggested in order to sustain the work generating ability of the fuel (Fuel Quality Report Cummins 2018). On the other hand, a minimum of 10% hydrogen addition in CNG was observed to be essential for considerable improvement in the performance characteristics (Sirens and Rosseel 2000). Thus, among the various compositions of HCNG blends tested as automotive fuel, 10% and 15% hydrogen in CNG were identified as the safest mixtures to conduct the performance and emission studies.

7.4 Engine Operation on Hydrogen-added Natural Gas

7.4.1 Method of fuel introduction

SI engines that are mostly made to run on gasoline need to be converted so that they can run on gaseous fuels. In this context, SI engines are classified as those in which the mixture of fuel and air goes into the engine cylinder. Accordingly, method and timing of induction for the mixture plays a critical role to determine combustion, performance & emission characteristic of the engine. Several fuel injection methods available as per the literature are presented in Table 7.2.

Among the tested methods, timed manifold injection (TMI) and direct cylinder injection were observed to be successful in achieving smooth engine operation using HANG blends. A detailed discussion on the comparison of various fuel injection systems can be found in the literature. However, timed manifold injection (TMI) was observed to be more appropriate than direct cylinder injection to ensure the safety of the engine unit. This system introduces the fuel at a definite point in the engine cycle. The ability of the injector to introduce the fuel effectively was tested using neat CNG at a supply pressure of 5 bars over the range of engine speed. Moreover, the performance of the system operating on timed manifold injection was observed to be quite satisfactory at a pressure of 2 kg/cm^2. Thus, the range of supply pressure remained usually on the lower side.

Table 7.2 Fuel injection methods (Das 1990).

Mixture	Classification	Flow timing of fuel	Supply pressure
Continuous carburation (CC)	Prior to inlet valve closure	Continuous flow	Minutely greater than atmospheric
Continuous manifold injection (CMI)	Prior to inlet valve closure	Continuous flow	Little greater than atmospheric
Timed manifold injection (TMI)	Prior to inlet valve closure	Hydrogen flow commences after opening of the intake valve but completed prior to intake valve closure	1.4–5.5 kg/cm^2
Low-pressure direct cylinder injection (LPDI)	Post inlet valve closure	Hydrogen flow commences after intake valve closure and is completed before significant compression pressure rise	2–8 kg/cm^2

7.4.2 Safety feature during engine operation

Hydrogen is considered a highly flammable and extremely explosive fuel. On a volumetric basis, hydrogen gas remains less explosive compared to methane or gasoline. Although, on a mass basis, it becomes altogether a different story (Table 7.1). It was mentioned earlier in the discussion that hydrogen is characterized by a high speed of flame propagation and a minimum ignition energy. Thus, relatively less time is required for the flame of hydrogen-mixed fuel to travel the distance. Moreover, such flames need very little energy to ignite rather than gasoline or methane. Consequently, a weak ignition source is observed to be sufficient to produce the spark. Therefore, safety measures during the combustion of hydrogen-mixed fuel in the engine system should be carefully considered. In this context, the installation of a non-return valve in the intake manifold is found to be valuable to avoid the backflow of fuel from the engine to the intake system. Flame traps that overcome possible burst can also be used on both sides of the possible ignition sources present at locations of the fuel flow channel. Likewise, leakage testing along with possible solutions to reduce wear and tear of parts should be performed in order to increase the reliability during the operation of combustion engines.

7.4.3 Phenomenon of undesirable combustion

A sufficient discussion on the benefits of adding hydrogen to compressed natural gas has been presented. Such gains can be obtained only if the stability in combustion necessary for smooth engine operation is achieved.

Accordingly, it is recommended that the chances of undesirable combustion (if any) produced due to the addition of hydrogen be rectified for smooth engine operation. Such phenomena may appear in various ways, like preignition, backfire, knocking, and a rapid rate of pressure rise. When the charge is ignited earlier than the actual timing, this is known as preignition. It is caused by the presence of hot spots. Intake and exhaust valves are well-known examples. It has been observed that when such combustion occurs close to the inlet valve, the flame may travel back toward the inlet manifold. Under severe conditions, it is quite possible that complete damage can happen to the induction system. Similarly, backfire occurs due to the burning of fuel during the intake stroke at the inlet manifold or inside the cylinder. Higher speed of flame propagation is one of the main reasons for the backfire phenomenon observed in hydrogen-fueled engines. On the other hand, there are chances that a knocking phenomenon may occur if backfire occurs near the exhaust valve. Knocking phenomenon is characterized by excessive intensity of high amplitude cylinder pressure oscillations. These oscillations occur due to the spontaneous burning of the fuel mixture present at the end of the engine cylinder. Consequently, a reduction in power output and the thermal efficiency of the engine can be noticed.

Several methods, such as exhaust gas recovery and water injection, were attempted in order to provide cooling of the hot spots, reducing the tendency for undesirable combustion. However, a timed manifold injection (TMI) also supported to avoid the tendency of preignition.

7.5 Performance and Emission Characteristics

The facility used for conducting the tests is schematically shown in Figure 7.2. The setup consists mainly of a single-cylinder gasoline engine coupled to an eddy-current type of dynamometer and a fuel supply arrangement. The tests were performed over a range of engine speeds (from 1000 to 2400 rpm) with original ignition timing and wide-open throttle (WOT) condition of the engine. A photograph of the actual spark-ignition engine coupled to an eddy current dynamometer is shown in Figure 7.3. The major specifications of the gasoline engine are shown in Table 7.3. Four different fuels, i.e., 10% and 15% HCNG mixtures, along with gasoline and pure methane, were considered for study. Gasoline and compressed natural gas were used to generate the baseline data in order to compare the engine characteristics obtained during HCNG operation.

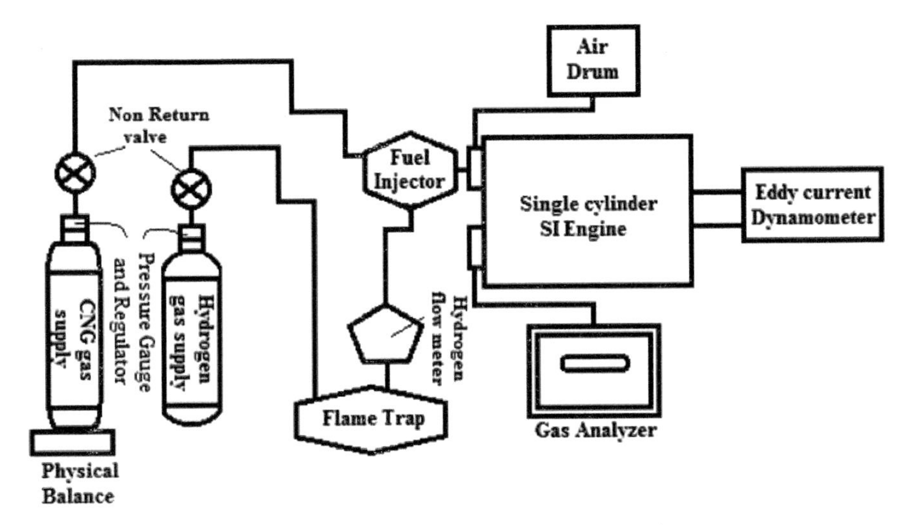

Figure 7.2 Schematic of the experimental test engine setup.

Table 7.3 Specifications of single-cylinder SI engine.

Parameter	Specifications
Bore × stroke	86 mm × 86 mm
Comp. ratio	10.5:1
Max. engine speed	3000 rpm
Air intake	Naturally aspirated
Injection system	Port fuel injection
Cooling system	Continuous water circulation

Gaseous fuels like CNG and mixtures of hydrogen and methane were introduced into the intake port by using a port injection system instead of the gasoline injector. This was done without making any changes to the gasoline injection system. Hydrogen and natural gas were supplied to the injector from the respective gas cylinders. A Lamda sensor was used on the engine to control the amount of fuel. The flow rate of natural gas was obtained by observing the weight of gas consumed over a period of time. Alternatively, a flow meter was utilized to determine the hydrogen flow rate. A flame trap and a non-return valve were also installed in the hydrogen supply line for safety during the test runs. An air drum method was used to determine the flow rate of air. An AVL Di Gas analyzer was used to measure the quantity of various pollutants, such as HC and NOx, delivered in the ambient.

Figure 7.3 Engine cylinder coupled to dynamometer.

7.5.1 Performance characteristics

The spark-ignition engines are designed and optimized to operate on gasoline. Thus, it may not be correct to make the comparison with gasoline fuel, but such a comparison will surely provide directions in the development of units capable of running on gaseous fuels.

Maximum power output depends on the method of fuel introduction. When the fuel is injected into the manifold, the gaseous fuel will displace the air. This results in lower volumetric efficiency and less power. On the other hand, spark timing becomes equally significant in determining the power developed as per the equivalence ratio. Spark timing needs to be advanced with a decrease in equivalence ratio. This is because the speed of flame propagation reduces at lower values of equivalence ratios.

7.5.1.1 Torque characteristics

An engine torque developed with the use of various fuels considered in the study is shown in Fig. 7.4. An increase in the value of torque with an increase in engine rpm was noticeable for all the fuels considered. The values of measured torque were observed to be better during the tests conducted with

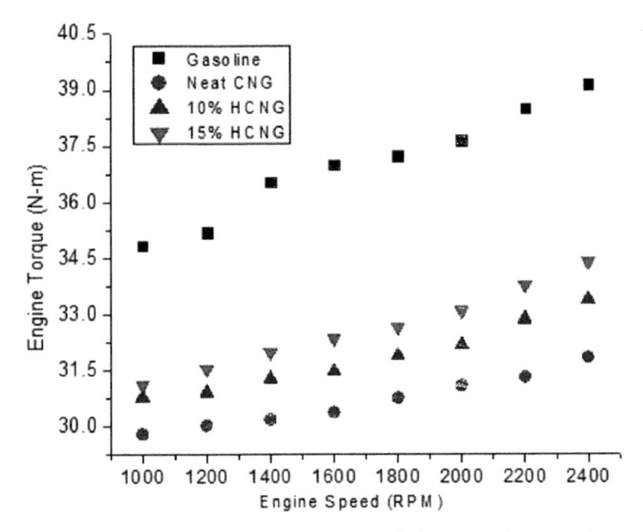

Figure 7.4 Engine torque characteristic vs. engine speed.

gasoline. A maximum value of 39.14 N-m was obtained at 2400 rpm. A reduction in this value by about 18%–14% for pure natural gas, 15%–12% for a 10% HCNG mixture, and 12%–10% for a 15% hydrogen natural gas mixture was recorded. The decrease in the value of torque produced may be attributed to the low energy density of gaseous fuels. Moreover, displacement of the air available for combustion by the gaseous fuels when introduced during the engine operation results in a reduction of volumetric efficiency.

On the other hand, improvement in produced torque was detected when compressed natural gas was replaced by mixtures of hydrogen added natural gas. An average gain of about 3.8% and 6.25% was obtained for experiments conducted with 10% HCNG and 15% HCNG, respectively, for the entire range of operation. This may be attributed to higher speed of flame propagation of hydrogen that helps in better rate of heat release for the new fuel formed by the addition of hydrogen. The enhancement in the values of torque obtained by adding small amount of hydrogen in CNG fueled engine is in agreement with the results obtained in many studies (Alrazen and Ahmed, 2018; Ji and Wang, 2011).

7.5.1.2 Thermal efficiency

The variation in brake thermal efficiency over entire range of engine speed for several fuels utilized during testing is shown in Fig. 7.5. The efficiency

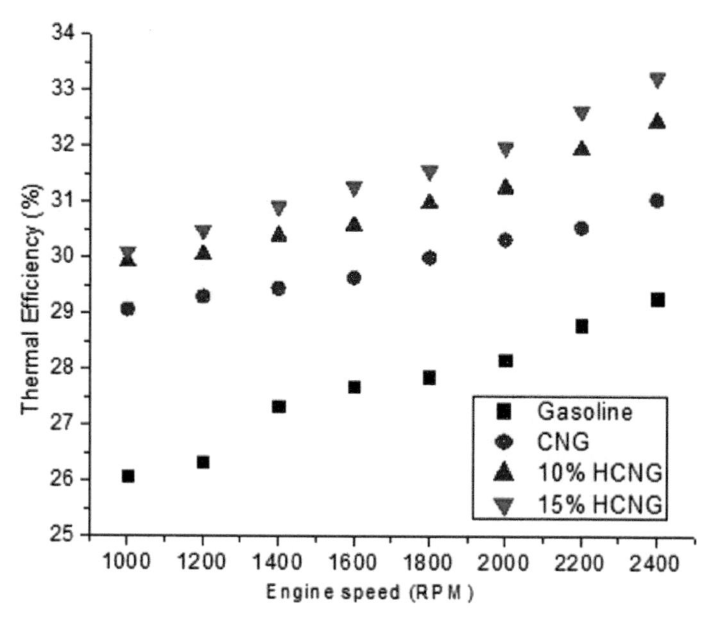

Figure 7.5 Thermal efficiency over the range of engine speed.

was observed to be minimum for tests conducted with conventional gasoline. An improvement in the values of brake thermal efficiency during CNG and HCNG operation was observed over the entire range of engine speeds. A mean increase in thermal efficiency of about 8.1% for neat CNG, 11.75% for 10% hydrogen in natural gas, and 13.8% for a 15% HCNG mixture was identified. This may be attributed to the behavior of gaseous fuels, which consume less energy for power production under similar conditions than gasoline. Moreover, hydrogen substitution in natural gas helps to achieve even better values of thermal efficiencies compared to pure natural gas due to the higher flame speed of hydrogen that results in rapid and sustained combustion phenomena. The results obtained are in agreement with the results reported in various other investigations (Ji et al., 2013; Dimopoulos et al., 2007).

7.5.1.3 Brake-specific energy consumption (BSEC)

Brake-specific energy consumption (BSEC) is a useful parameter that is used to compare fuels with different Energy densities. The energy requirements of various fuels, such as pure gasoline, clean natural gas, 10% hydrogen in natural gas, and a 15% HCNG mixture is presented in Figure 7.6. A

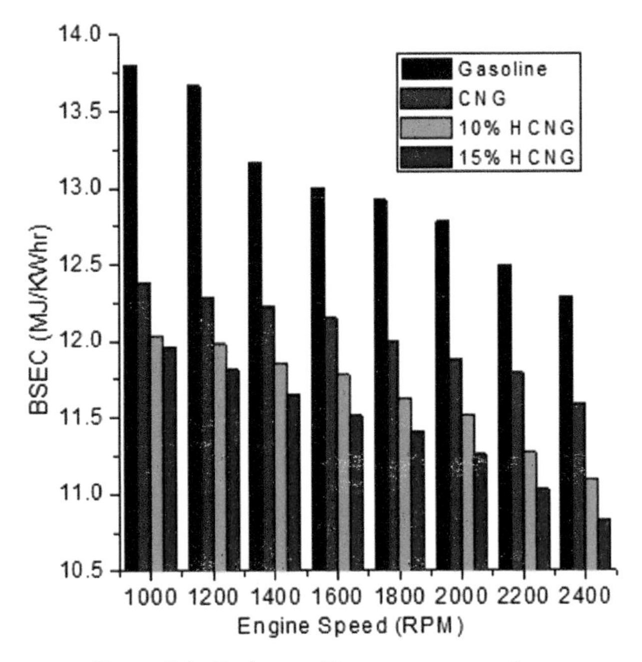

Figure 7.6 Brake-specific energy consumption.

maximum amount of energy consumption was observed during engine tests operated with gasoline. A decrease in this value was noticed when the fuel was changed from gasoline to neat CNG and then to HCNG mixtures. An average reduction of about 7.4% for natural gas, 10.5% for 10% hydrogen in natural gas, and 12.1% for 15% hydrogen in natural gas was identified during the experimental runs.

7.5.2 Emission characteristics

Various pollutants, such as HC, CO, and CO_2 particulates, are observed to decrease with an increase in hydrogen percentage. Even the traces of aromatics and aldehydes in the exhaust were not observed in the tests conducted with neat hydrogen. However, the presence of NOx in the exhaust of the engine running on pure or blends of hydrogen has been observed (Bauer and Forest, 2001; Zhao et al., 2013; Bedi, 2020).

The amount of NOx in the exhaust is only a concern for an automotive engine operated on hydrogen blends. The ill effects of nitrogen oxides include ozone formation and respiratory problems. The temperature of the combustion chamber and the availability of oxygen are the two factors that play a key role in determining the amount of NOx emissions. This indicates that the change in equivalence ratio can regulate nitrogen oxides at the outlet. Conducting the engine experiments with rich mixtures will reduce the amount of oxygen, but this will increase the chances of undesirable combustion, such as backfiring. The technique of exhaust gas recovery is also observed to be effective in the control of NOx emissions. On the other hand, lean mixtures result in lower temperatures, which in turn result in lesser NOx emissions.

7.5.2.1 Hydrocarbon emissions

The variation in HC emissions obtained over the entire range of engine speed for the different fuels considered under testing is presented in Figure 7.7. The maximum amount of hydrocarbons can be noticed during gasoline operation. Incomplete combustion, along with the wall quenching effect, was supposed to be responsible for its generation. A reduction in these emissions was

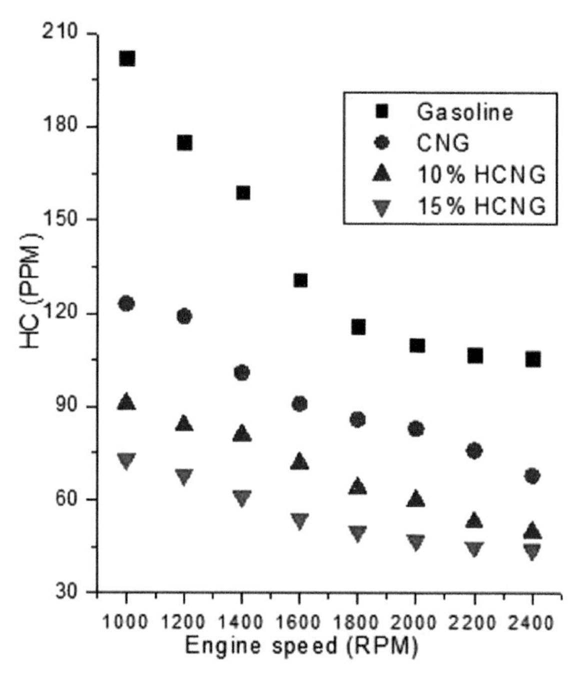

Figure 7.7 Hydrocarbon emissions vs. engine speed.

observed when engines were made to operate on gaseous fuels. Hydrocarbon emissions were observed to be minimal for the fuel obtained by the substitution of 15% hydrogen in CNG. This may be attributed to the decrease in methane content of the fuel obtained by adding hydrogen. An average reduction of about 25.8% and 40.6% in HC emissions was observed when engines were made to operate on 10 HCNG and 15 HCNG mixtures rather than neat compressed natural gas. On the other hand, a decrease in the amount of hydrocarbon emissions can be observed at higher speeds of the engine. This may be due to the excessive temperatures achieved within the cylinder due to lower heat losses noticed at higher engine speeds that result in complete combustion.

7.5.2.2 Nitrogen oxides emissions

At the high temperatures found inside the engine cylinder during combustion, the nitrogen gas in the air tends to mix with the oxygen. So, high temperature along with presence of oxygen in adequate quantity was observed to favour the formation of NOx emissions. Figure 7.8 shows the amount of nitrogen

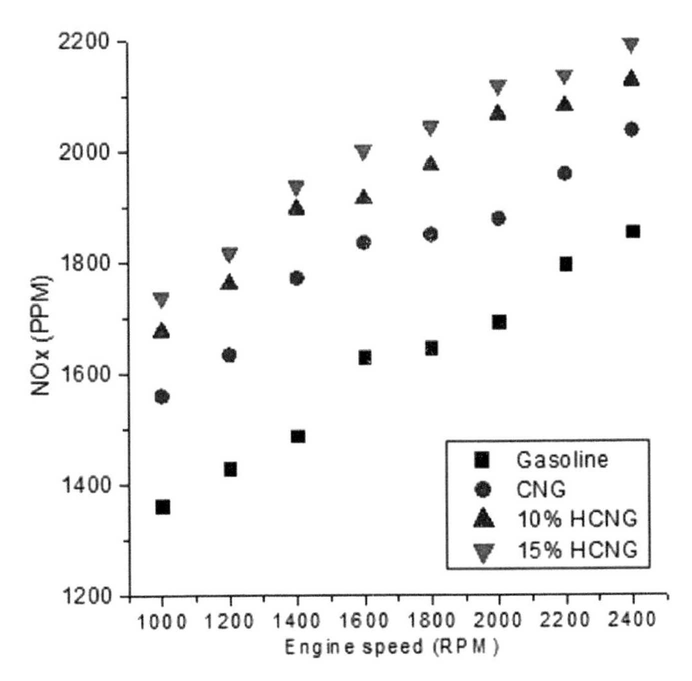

Figure 7.8 Nitrogen oxides vs. engine speed.

oxides produced in the exhaust when the engine was tested with pure gasoline, natural gas, 10% HCNG, and 15% HCNG. The amount of Nitrogen oxides were found to be produced the least when gasoline was used, a bit more when CNG was used, and the most during engine operation on natural gas with 15% hydrogen. When the fuel was changed from CNG to 10% HCNG and then to 15% HCNG, the average amount of nitrogen oxides in the exhaust went up by 6.7% and 10.1%, respectively. So, hydrogen addition helps in producing higher amount of NOx by increasing the temperature inside the cylinder. Enhancement in nitrogen oxide emissions with hydrogen substitution in the utilized fuel was also reported in many other studies as well (Deng et al., 2011; Mariani et al., 2012; Tinaut et al., 2011).

7.6 Concluding Remarks

The studies conducted on the performance and emissions of hydrogen-fueled engines indicated an improvement in engine performance with the addition of hydrogen to CNG over neat CNG operation. Better torque and brake thermal efficiency of the engine were observed. A decrease in specific energy consumption was also noticed. However, under similar conditions, the torque produced was observed to be lower than the torque produced with the use of gasoline. Thus, the use of techniques in order to compensate for the loss of work is recommended for the implementation of HCNG-fueled engines. On the other hand, improvement in hydrocarbon emissions was also recorded with the use of HCNG mixtures due to a lower quenching distance and an increase in the percentage of hydrogen. On the other hand, higher amounts of NOx emissions with a greater amount of hydrogen in the mixture are a matter of concern. Consequently, EGR or a catalytic converter must be used to control the amount of NOx produced in HCNG-fueled engines.

Acknowledgement

This chapter is based on the work conducted at the Engines and Unconventional Fuels laboratory in the Centre for Energy Studies, IIT Delhi under the guidance of Professor L.M. Das and did not receive any grant from anywhere else. I am thankful to him for his guidance and encouragement while conducting the work.

References

[1] Acıkgoz BCÿ Celik C, Soyhan HS, Gokalp B, Karabag B. (2015) Emission characteristics of hydrogen- CH4 fueled spark ignition engine. Fuel 159, 298-307. DOI 10.1016/j.fuel.2015.06.043

[2] Akansu SO, Kahraman N, Çeper B. (2007) Experimental study on SI engine fueled by methane–hydrogen mixtures. Int J Hydrogen Energy 32, 4279–84. DOI 10.1016/j.ijhydene.2007.05.034

[3] Alrazen H A, Ahmad K A, (2018) HCNG fueled spark-ignition engine with its effects on performance and emissions, Renewable and Sustainable Energy Reviews 82, 324–342. DOI 10.1016/j.rser.2017.09.035

[4] Bauer C G, Forest T W, (2001) Effect of hydrogen addition on the performance of methane-fueled vehicles. Part I: effect on S.I. engine performance. Int J Hydrogen Energy 26, 55–70. DOI 10.1016/S0360-3199(00)00067-7

[5] Bedi N, (2020) Utilization of Hydrogen-CNG Blends in a port injected Spark Ignition Engine, In Proceedings of 8th International and 47th National Conference on Fluid Mechanics and Fluid Power (FMFP) December 09-11, IIT Guwahati.

[6] Biffiger H, Soltic P, (2015) Effects of split port/direct injection of methane and hydrogen in a SI engine, Int J Hydrogen Energy 40, 1994–2003. DOI 10.1016/j.ijhydene.2014.11.122

[7] Ceper BA, Akansu SO, Kahraman N, (2009) Investigation of cylinder pressure for H_2/CH_4 mixtures at different loads. Int J Hydrogen Energy 34, 4855–61.

[8] DOI 10.1016/j.ijhydene.2009.03.039

[9] Colin R F, Allen T. (2001) Internal combustion engine. New York: John Wiley & Sons.

[10] Das, L M, (1990) Fuel induction techniques for hydrogen operated engine. Int. J. Hydrogen Energy 15, 833–842. DOI 10.1016/0360-3199(90)90020-y

[11] Das L M, (1996) Hydrogen-oxygen reaction mechanism and its implication to hydrogen engine combustion. Int J Hydrogen Energy 21, 703–15. DOI 10.1016/0360-3199(95)00138-7

[12] Das L M, (1996) Utilization of hydrogen - CNG blends in an internal combustion engine' In Proceedings of 11th World hydrogen energy conference, Stuttgart, Germany. 23-28 1513- 1535.

[13] Das L M and Lather R S, (2019) Performance and emission assessment of a multi cylinder S.I engine using CNG & HCNG as fuels. Int J Hydrogen Energy 44, 21181-21192. DOI 10.1016/j.ijhydene.2019.03.137

[14] Das L M, Gulati R, Gupta P K, (2000) A comparative evaluation of the performance characteristics of a spark ignition engine using hydrogen and compressed natural gas as alternative fuels. Int J Hydrogen Energy 25, 783 – 793. DOI 10.1016/s0360-3199(99)00103-2

[15] Deng J, Ma F, Li S, He Y, Wang M, Jiang L, Zhao S(2011) Experimental study on combustion and emission characteristics of a hydrogen-enriched compressed natural gas engine under idling conditions. Int J Hydrogen Energy 36, 13150–57. DOI 10.1016/j.ijhydene.2011.07.036

[16] Dimopoulos P, Rechsteiner C, Soltic P, Laemmle C, Boulouchos K, (2007) Increase of passenger car engine efficiency with low engine-out emissions using hydrogen– natural gas mixtures: a thermodynamic analysis, Int J Hydrogen Energy 32, 3073–83. DOI 10.1016/j.ijhydene.2006.12.026

[17] Eichlseder H, Klell M. Wasserstoff in der Fahrzeugtechnik [Hydrogen in Vehicle Technology]. 2nd ed. Wiesbaden, ISBN: ViewegþTeubner; 2010. 9783834810274.

[18] Eichlseder H, Wallner T, Freymann R, Ringler J, (2003) The potential of hydrogen internal combustion engines in a future mobility scenario, SAE, Technical Paper No. 2003-01-2267. DOI 10.4271/2003-01-2267

[19] Fuel Quality Calculator. Cummins westport. 2018,

[20] https://www.cumminswestport.com/fuel-qualitycalculator

[21] Hora T S, Agarwal A K, (2015) Experimental study of the composition of hydrogen enriched compressed natural gas on engine performance, combustion and emission characteristics. Fuel 160, 470–78. DOI 10.1016/j.fuel.2015.07.078

[22] Ji C, Wang S, (2011) Effect of hydrogen addition on lean burn performance of a spark-ignited gasoline engine at 800 rpm and low loads. Fuel 90, 1301–1304. DOI 10.1016/j.fuel.2010.11.014

[23] Ji C, Zhang B, Wang S, (2013) Enhancing the performance of a spark-ignition methanol engine with hydrogen addition. Int J Hydrogen Energy 38, 7490–98. DOI 0.1016/j.ijhydene.2013.04.001

[24] Kahraman E, Ozcanlı S C, Ozerdem B, (2007) An experimental study on performance and emission characteristics of a hydrogen fueled spark ignition engine, Int J Hydrogen Energy. 32, 2066–72. DOI 10.1016/j.ijhydene.2006.08.023

[25] Kamil M, Rahman M. M, (2015) Performance prediction of spark-ignition engine running on gasoline-hydrogen and methane-hydrogen blends, Applied Energy. 158, 556–567. DOI 10.1016/j.apenergy.2015.08.041

[26] Klell M, Eichlseder H, Sartory M, (2012) Mixtures of hydrogen and methane in the internal combustion engine – synergies, potential and regulations, Int J Hydrogen Energy. 37, 11531–40. DOI 10.1016/j.ijhydene.2012.03.067

[27] Laget O, Richard S, Serrano D, Soleri D, (2012) Combining experimental and numerical investigations to explore the potential of downsized engines operating with methane/hydrogen blends, Int J Hydrogen Energy. 37, 11514–30. DOI 10.1016/j.ijhydene.2012.03.153

[28] Mariani A, Morrone B, Unich A, (2012) Numerical evaluation of internal combustion spark ignition engines performance fueled with hydrogen–natural gas blends, Int J Hydrogen Energy. 37, 2644–54. DOI 10.1016/j.ijhydene.2011.10.082

[29] Mehra R. K., Duan H, Juknelevicius R, Ma F, Li J, (2017) Progress in Hydrogen enriched compressed natural gas internal combustion engines, Renewable and sustainable energy reviews. 80, 1458 – 1498. DOI 10.1016/j.rser.2017.05.061

[30] Moreno F, Munoz M, Arroyo J, Magen O, Monne C, Suelves I, (2012) Efficiency and emissions in a vehicle spark ignition engine fueled with hydrogen and methane blends, Int J Hydrogen Energy. 37, 11495 – 11503. DOI 10.1016/j.ijhydene.2012.04.012

[31] Renny A and Janardan S, (2008) Hydrogen - CNG blend performance in a three-wheeler. SAE, Technical Paper No. 2008-28-0119. DOI

[32] Saravanan N, Nagarajan G, Dhanasekaran C, Kalaiselvan K, (2007) Experimental investigation of hydrogen port fuel injection in DI diesel engine, Int J Hydrogen Energy. 32, 4071–80. DOI 10.1016/j.ijhydene.2007.03.036

[33] Sirens R, Rosseel E, (2000) Variable composition hydrogen/natural gas mixtures for increased engine efficiency and decreased emissions, J Engg Gas Turbines Power. 122, 135- 140. DOI 10.1115/1.483191

[34] Subramanian M, (2014) Performance analysis of 18% HCNG fuel on heavy-duty engine. SAE. Technical Paper. No. 2014-01-1453. DOI 10.4271/2014-01-1453

[35] Tangoz S, Kahraman N, Akansu SO, (2017) The effect of hydrogen on the performance and emission of an SI engine having compression ratio

fueled by compressed natural gas. Int J of Hydrogen Energy. 42, 1-15. DOI: 10.1016/j.ijhydene.2017.04.076

[36] Thipse S, Rairikar S, Kavathekar K, Chitnis P, (2009) Development of a six cylinder HCNG engine using an optimized lean burn concept. SAE. Technical Paper No. 2009-26-0031. DOI 10.4271/2009-26-0031

[37] Tinaut F, Melgar A, Giménez B, Reyes M, (2011) Prediction of performance and emissions of an engine fueled with natural gas/hydrogen blends, Int J Hydrogen Energy. 36, 947–56. DOI 10.1016/j.ijhydene.2010.10.025

[38] Wang S, Ji C, Zhang J, Zhang B, (2011) Comparison of the performance of a spark-ignited gasoline engine blended with hydrogen and hydrogen–oxygen mixtures, Energy. 36, 5832–37. DOI 10.1016/j.energy.2011.08.042

[39] White C, Steeper R, Lutz A, (2006) The hydrogen-fueled internal combustion engine: a technical review. Int J Hydrogen Energy. 31, 1292–305. DOI 10.1016/j.ijhydene.2005.12.001

[40] Zhao J, Ma F, Xiong X, Deng J, Wang L, Naeve N, Zhao S. (2013) Effects of compression ratio on the combustion and emission of a hydrogen enriched natural gas engine under different excess air ratio. Energy. 59, 658–65. DOI 10.1016/j.energy.2013.07.033

8

Performance Analysis of Hydrogen as a Fuel for Power Generation

Karthik Kumar, Meeta Sharma, and Anoop Kumar Shukla

Department of Mechanical Engineering, Amity University, India
E-mail: karthikkumar131099@gmail.com, msharma15@amity.edu
shukla.anoophbti@gmail.com

Abstract

The amount of pollution in the environment has gotten much worse as people have used more conventional fuels and released more pollution into the air. This has led to a search for clean and energy-efficient sources. The main factors to be considered are emissions and sustainability for future uses. Hydrogen is a strong contender in this arena since it is available in abundance and can be obtained through the electrolysis of water or digestion of hydrocarbons. Hydrogen is already being used in refineries, but for power generation, its usage has not been analyzed. Hydrogen can play a major role as an energy-efficient fuel in gas turbines. This chapter focuses on the performance analysis of a gas turbine power plant system built based on the first and second laws of thermodynamics. Analysis results indicate that amalgamating hydrogen and natural gas can benefit the system, and increasing the concentration of said hydrogen will positively impact the system. The effect of steam injection along with hydrogen has also been investigated, and the effect of using blended fuel and steam injection has also been analyzed and discussed. Thermodynamic analyses have been performed on the system, and the individual components, energy degradation, and efficiency have been evaluated. Energy degradation has been noticed to be the highest at the combustor, and, consequently, the efficiency is the lowest, with an average value of 55%–56%. The air compressor efficiency has been observed to be 94.14%, and, hence, the energy degradation in the air combustor is the least. The addition of hydrogen and steam to natural gas has increased the work output of the system by an average of 10%, depending on the proportion of hydrogen and steam used. The results of all the investigative analysis for all the operating conditions will be scrutinized in the forthcoming sections.

Keywords: Gas turbine, power plant, hydrogen fuel, power generation

Nomenclature

T_0	Dead-state temperature, K
T_1	Compressor inlet temperature, K
T_2	Compressor outlet temperature, K
T_3	Turbine inlet temperature, K
T_4	Turbine outlet temperature, K
P_0	Dead-state pressure, kPa
P_1	Compressor inlet pressure, kPa
P_2	Compressor outlet pressure, kPa
P_3	Turbine inlet pressure, kPa
P_4	Turbine outlet pressure, kPa
R	Pressure ratio
¥	Specific heat ratio, J/ kg K
W_C	Compressor work, kW
W_{GT}	Gas turbine work, kW
W_{netGT}	Network, kW
$C_{p\ air}$	Specific heat capacity of air, kJ/kg K
$C_{p\ gases}$	Specific heat capacity of gas, kJ/kg K
m_{air}	Mass flow rate of air, kg/s
m_{gas}	Mass flow rate of the gases, kg/s
PLC	Pressure loss constant
m_{CH4}	Mass flow rate of natural gas, kg/s
m_{H2}	Mass flow rate of hydrogen gas, kg/s
m_{H2O}	Mass flow rate of steam, kg/s
$\dot{\eta}$	Overall system efficiency
EX_1	Compressor inlet exergy flow, kW
EX_2	Compressor outlet exergy flow, kW
EX_3	Turbine inlet exergy flow, kW
EX_4	Turbine outlet exergy flow, kW
$EX_{physical}$	Physical exergy, kW
EX_{chem}	Chemical exergy, kW
EX_P	Exergy due to pressure, kW
EX_T	Exergy due to temperature, kW
R	Gas constant, J/k mol
EX_D	Exergy destruction, kW
$\dot{\eta}_{ex}$	Exergy efficiency

8.1 Introduction

Many countries across the world have started moving toward clean and alternate energy sources after witnessing firsthand the effects of increased dependence on fossil fuels. These alternate energy sources are sustainable and can play a major role in developing and transforming the schematics of various industries across the globe, and the reduced footprint with regard to carbon and other emissions is an additional bonus. In terms of the broad picture, the overall effect of these energy sources is positive, and a lot of industries have already started using these alternate energy sources. Hydrogen is one such element that has witnessed an exponential rise in its production and usage since the search for an alternate source started. Hydrogen has a lot going for it, right from the ease of obtaining it to the lightness of its molecule. The environmental institution of California has assessed this and reported that in the future, hydrogen will have the capability to replace fossil fuel in every sense. The rapid pace at which energy sources are being depleted has prompted researchers in the scientific community to figure out ways to increase the efficiency of power generation plants. One of the vital tools that will help the researchers with this is the law of thermodynamics. An analysis of the energy aspect will depict the efficiency of the overall system, but performing the exergy analysis of the system will pave the way for understanding the problems and the possible solutions. Exergy analysis will also help in understanding the energy degradation in each component and possible ways to decrease it, which will consequently increase efficiency. Gas turbines can offer a flexible amount of power for generation, and for this purpose, gas turbines mainly use natural gas. While natural gas emits lower amounts of NOx and carbon compounds, the emissions need to be controlled more in order to reach the levels stated in the Paris Agreement, which primarily mandates carbon neutrality. This fact is made clearer by a statement issued by the European Investment Bank, which states that the bank will no longer fund projects that will run on fossil fuels with effect from 2021. According to this statement, power generation plants that exceed 250 g CO_2/kWh_{el} will no longer receive any funding. This limit of 250 g CO_2/kWh_{el} completely eliminates the use of coal as a fuel and the use of natural gas fired gas turbines since these gas turbines would have an emission of about 330 g CO_2/kWh_{el} assuming 60% efficiency. Further reduction of carbon emissions from these gas turbines is only possible in either of the two following ways:

- switching to fuel with better properties and lesser emissions;
- improving the thermal efficiency.

The latter will only marginally contribute toward emission reduction, as can be understood by thermodynamics, and this leaves us to go with the better option, which is switching to a more efficient fuel. Many different fuels can be considered as alternatives, and gas turbines can also use a variety of blends [1, 2]. In recent times, the role and importance of hydrogen have increased to a great extent due to its potential as a viable and feasible energy source 3–9. The usage of hydrogen gas to power gas turbines was first experimented with by Bothien et al. [10], and they found out that the carbon emissions were lowered to a great extent while using hydrogen. For this purpose, they used an Ansaldo Energia sequential combustion gas turbine. The economic aspects of utilizing hydrogen to power gas turbines were considered by Gibrael and Hassan [11] and they concluded that while using natural gas to power gas turbines was more cost-efficient in today's time, the cost effectiveness of the said hydrogen gas will be much better than that of natural gas in the upcoming future. Many analytical, digital, and empirical studies were conducted by Bozo et al. [12] to determine the efficiency of hydrogen combined with ammonia, and the use of empirical investigation using steam injection showcased a massive jump of 30% in terms of productivity of the entire system. The world's first ever fully hydrogen powered gas turbine was developed in Japan by Kawasaki, Tekin, et al. [13].

In the present study, investigation is performed on the use of hydrogen blended with natural gas and steam in varying proportions to understand the effect of blended fuel (natural gas, hydrogen, and steam) in a gas turbine. The use of only hydrogen fuel to power gas turbines has not been investigated, but, rather, the use of hydrogen in a certain range of proportion has only been investigated because of the aforementioned cons of hydrogen and because, with the current technology of gas turbines, shifting completely toward hydrogen will prove to be a costly affair. For now, blending hydrogen with conventional fuels like natural gas will rather be beneficial since the system does not need to undergo any major changes.

8.2 System Layout and Description

Figures 8.1 and 8.2 show the layout that was made for simulation. This layout was made using the laws of thermodynamics. The main components of the system include the air compressor, combustor, and turbine. The workings of the system can be explained as follows:

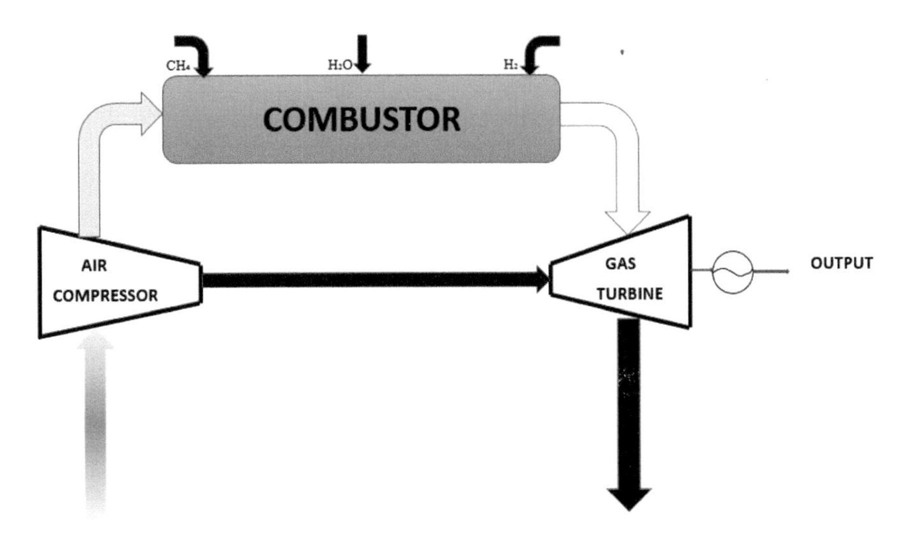

Figure 8.1 Layout of non-recuperative gas turbine system.

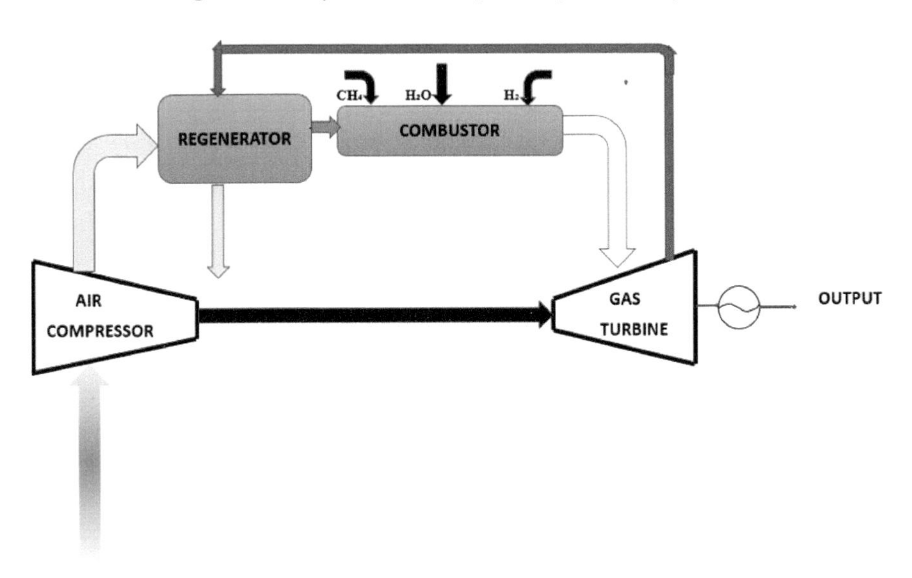

Figure 8.2 Recuperative gas turbine system layout.

- The compressor takes in air and then squeezes it to make it hotter and more pressurized before sending it to the combustion chamber to mix with the fuel.

- Different amounts of hydrogen, natural gas, and steam are mixed to make the fuel in question. The fuel combusts after being injected into the combustion chamber, and the high-temperature fuel is sent to the turbine, where it expands and produces work.
- In the presence of a heat exchanger in the form of a regenerator, the compressed air gets pre-heated through the regenerator, and this pre-heated air is sent over to the combustion chamber.
- In the combustor, the fuel is injected and ignited.
- Since the amount of heat in the air is greater, ignition occurs more easily, and the network and efficiency both increase.

8.3 Mathematical Modeling

8.3.1 The proposed layout of the system and the mathematical model of the system

The model developed consists of the following major components: compressor, combustion chamber, gas turbine, and regenerator.

8.3.1.1 Compressor model

The outlet temperature of the compressor can be calculated using the following formula:

$$T_2 = T_1 \left[\left(1 + \frac{1}{\eta_c \left(r^{\left(\frac{\gamma-1}{\gamma}\right)} - 1 \right)} \right) \right] \tag{8.1}$$

where r is the compressor pressure ratio, and γ is the specific heats ratio.

The specific heat capacity of air, C_p, can be calculated from the following formula:

$$C_{p\ \text{air}} = 1.048 - \left(\frac{3.83T}{10^4}\right) + \left(\frac{9.45T^2}{10^7}\right) - \left(\frac{5.49T^3}{10^{10}}\right) + \left(\frac{7.94T^4}{10^{14}}\right). \tag{8.2}$$

Air compressor work rate can be calculated as follows:

$$W_C = m_{\text{air}} C_{p\ \text{air}} (T_2 - T_1) \tag{8.3}$$

where W_C is the work done by the compressor, m_{air} = mass of the air, $C_{p\ \text{air}}$ = specific heat capacity of air, and T_1 and T_2 are the inlet and outlet temperatures.

8.3.1.2 Regenerator model

In the case of regenerator, the effectiveness of the regenerator can be calculated through the following expression:

$$€_{\mathrm{Regenerator}} = \frac{T_{2'} - T_2}{T_4 - T_2} \tag{8.4}$$

where $€_{\mathrm{Regenerator}}$ is the effectiveness of the regenerator, which is assumed to be 95%.

8.3.1.3 Combustion chamber model

In case of the combustion chamber, we are assuming a constant pressure loss of 5% and if PLC is the pressure loss constant

$$\frac{P_s}{P_2} = 1 - PL_{\mathrm{cc}}. \tag{8.5}$$

The mass balance equation for the combustion chamber can be written down as follows:

$$m_{\mathrm{gases}} = m_{\mathrm{air}} + m_{\mathrm{CH4}} + m_{\mathrm{H2}} + m_{\mathrm{H20}}. \tag{8.6}$$

The energy balance equation for the combustion chamber can be expressed as follows:

$$\begin{aligned} m_{\mathrm{air}} \cdot C_{\mathrm{pair}} \cdot T_2 + Q_{\mathrm{H_2}} + Q_{\mathrm{CH_4}} + Q_{\mathrm{H_2O}} \\ = (m_{\mathrm{air}} + m_{\mathrm{H_2}} + m_{\mathrm{CH_4}} + m_{\mathrm{H_2O}})\ C_{p\ \mathrm{gas}} \cdot T_3 \end{aligned} \tag{8.7}$$

The specific heat capacity of gases, C_p, can be calculated using the following equation:

$$C_{p\ \mathrm{gases}} = 0.991 - \left(\frac{6.997T}{10^5}\right) + \left(\frac{2.712T^2}{10^7}\right) - \left(\frac{1.2244T^3}{10^{10}}\right). \tag{8.8}$$

8.3.1.4 Turbine model

In case of the turbine, the work produced can be evaluated from

$$W_{\mathrm{GT}} = m_{\mathrm{gases}} * C_{p\ \mathrm{gases}} * (T_3 - T_4) \tag{8.9}$$

where W_{GT} is the work produced by the gas turbine, $C_{p\ \mathrm{gases}}$ is the specific heat capacity of the gases, and T_3 and T_4 are the temperature at the inlet and outlet of the gas turbine.

The amount of net-work produced by the system can be evaluated from the following:

$$W_{\mathrm{Net}} = W_{\mathrm{GT}} - W_{\mathrm{Compressor}}. \tag{8.10}$$

The overall efficiency of the system can be calculated from the following equation:

$$\eta = W_{\mathrm{Net}}/Q_{\mathrm{system}}. \tag{8.11}$$

8.3.2 Exergy model

Exergy can be defined as the maximum possible amount of work that a system can produce with reference to the dead-state condition. The phenomenon of exergy occurs due to its irreversibility in the system. Exergy in a gas turbine system consists of four main components, which are physical, chemical, potential, and kinetic exergy. Physical exergy occurs due to the deviation of the system's temperature and pressure from the environment, chemical exergy occurs due to the deviation in the system's chemical composition with respect to the environment, kinetic exergy is due to the system's velocity compared relatively with the environment, and potential exergy is due to the system's height compared relatively with the environment. In the case of gas turbine power plants, the kinetic and potential exergy can be neglected as they are too small to be considered.

Physical exergy: Physical exergy occurs due to the deviation of the system's temperature and pressure from the environmental conditions. Physical exergy can be calculated from the following equations:

$$\mathrm{EX}_{\mathrm{physical}} = \mathrm{EX}_{T} + \mathrm{EX}_{P} \tag{8.12}$$

$$\mathrm{EX}_{T} = C_p\left((T - T_0) - T_0 \cdot \ln\left(\frac{T}{T0}\right)\right) \tag{8.13}$$

$$\mathrm{EX}_{P} = R \cdot T_0 \cdot \ln\left(\frac{P}{P_0}\right). \tag{8.14}$$

Physical exergy can be further divided into exergy due to pressure and exergy due to temperature and these can be calculated from eqn (8.14) and (8.15). T_0 and P_0 denote the dead-state temperature and pressure, while R and C_p denote the gas constant and the specific heat due to constant pressure, respectively.

Table 8.1 Standard molar chemical exergy values [3, 14].

Substance	e_{xi}^{ch}**(kJ/mol)**
H_2O	9.49
N_2	0.72
O_2	3.97
H_2	236.1
CO_2	19.87

Chemical exergy: Chemical exergy occurs due to the change in the chemical composition of the system from the environmental conditions and this can be calculated from the following equations:

$$\pounds = \frac{\mathrm{ex_{fuel}}}{\mathrm{LHV}}. \tag{8.15}$$

In eqn (8.16), the £ sign represents the ratio of exergy flow of the fuel to the lower heating value (LHV) of the fuel. The ratio of exergy flow to the LHV is usually taken as 1.06 in case of natural gas. Through this equation, the exergy flow of the fuel can be determined.

For calculating the exergy of the product of the combustion gases (as given in the Table 8.1), the following equation may be used:

$$\mathrm{EX_{chemical}} = \Sigma x_k \cdot ex_{\mathrm{ch}}^k + RT_0 \Sigma\, x_k \cdot \ln\left(x_k\right). \tag{8.16}$$

In eqn (8.16), x_k denotes the molar fraction of the gas, R denotes the universal gas constant, T_0 is the ambient temperature, and ex_{ch}^k depicts the conventional value of molar chemical exergy of the gas.

The overall exergy can be calculated for all the components of the system using the following equation:

$$\mathrm{EX} = \mathrm{EX_{physical}} + \mathrm{EX_{chemical}} \tag{8.17}$$

Exergy destruction: The exergy flow rate calculated for every individual component can be used to evaluate the exergy destruction. After each process, there is a loss of exergy and this loss is known as exergy destruction. The exergy destruction for each component can be determined from the following equations.

Air compressor efficiency:

$$\mathrm{EX_1} + W_{compressor} = \mathrm{EX_2} + \mathrm{EX}_D. \tag{8.18}$$

Combustion chamber efficiency:

$$\mathrm{EX_2} + \mathrm{EX_5} = \mathrm{EX_3} + \mathrm{EX}_D. \tag{8.19}$$

Gas turbine efficiency:

$$\mathrm{EX}_3 = \mathrm{EX}_4 + W_{GT} + \mathrm{EX}_D. \tag{8.20}$$

Performance analysis:

The exergy flow rate for every component and the associated exergy destruction can be calculated from eqn (8.13) to eqn (8.19). Hence, the performance analysis of the gas turbine system can be performed using the first and second laws of thermodynamics. Through this analysis, the performance of the individual components as well as the entire system can be determined and evaluated.

Air compressor:

$$\dot{\eta}_{\mathrm{AC}} = 1 - \left(\mathrm{EX}_D / W_{\mathrm{compressor}} + \mathrm{EX}_1\right). \tag{8.21}$$

Combustion chamber:

$$\dot{\eta}_{CC} = 1 - \left(\mathrm{EX}_D / \mathrm{EX}_2 + \mathrm{EX}_3 + \mathrm{EX}_5\right). \tag{8.22}$$

Gas turbine:

$$\dot{\eta}_{\mathrm{GT}} = 1 - \left(\mathrm{EX}_D / W_{\mathrm{GT}} + \mathrm{EX}_3\right). \tag{8.23}$$

8.4 Results and Analysis

The model for the entire system and scenarios were designed using C++ software. The fuel used in the system is a blend of natural gas, hydrogen, and steam in varying proportions since hydrogen is widely considered a clean and very challenging fuel of the future. Due to hydrogen's high energy density and efficiency, its effect on power generation with respect to gas turbines has been investigated. The specific properties of hydrogen make it exciting and demanding for its future prospects. Since hydrogen is costly, using hydrogen along with conventional natural gas has been analyzed, and the effect of injecting steam in the amalgamation has also been discussed to evaluate the change in performance values and efficiency. The results obtained for all the scenarios were analyzed for a variety of operating conditions and are reviewed in the following section.

Scenario 1: Effect of 95% natural gas and 5% hydrogen fuel:

Figures 8.3 (a) and 8.3 (b) depict the variation in the network at different pressure ratios and turbine inlet temperatures for a non-recuperative and recuperative system, respectively. Figure 8.3 (a) ascertains that as temperature increases, the value of the network increases and subsequently decreases with increasing pressure ratio. According to the equation, it can be deduced that when pressure is constant, temperature and volume are directly proportional, and when temperature is constant, pressure and volume are inversely proportional. When temperature increases, the energy of the individual atoms increases, and this temperature change causes thermal expansion. Gases are majorly affected by this thermal expansion, and since the fuel used is essentially a mixture of gases, the temperature change causes a drastic thermal expansion of the gaseous atoms, which in turn produces more work. When the temperature is constant, the pressure is inversely proportional to the volume of gas and consequently the temperature, and this explains why the amount of net-work decreases with increasing pressure ratio at a constant inlet temperature. Figure 8.5 (b) depicts the variation in net-work at different pressure ratios and temperatures, but, here, a regenerator is used to observe any variation in net-work produced. It is ascertained that the amount of net-work is slightly higher in this case, and this can be explained by the fact that the regenerator uses some of the leftover energy from the previous cycle of operation to pre-heat the compressed air before it makes its way over to the combustion chamber. This process helps the air get heated to a relatively higher value, which affects the amount of net-work produced. From the above figures, the change can be noticeably observed, and this change consequently affects the system's efficiency as well. From Figures 8.4 (a) and 8.4(b), the change in efficiency can be noticeably observed, and this change occurs due to the presence of the regenerator. Since a lesser amount of heat is needed in the case of a regenerator, the amount of net-work increases, increasing the

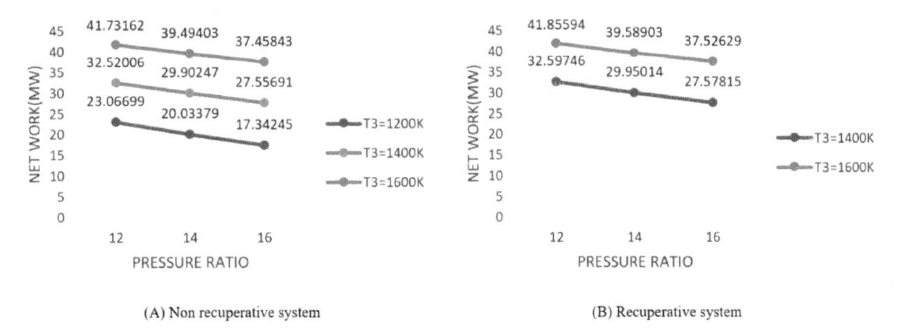

Figure 8.3 Variation in network with pressure ratio for blend of 95% natural gas and 5% hydrogen.

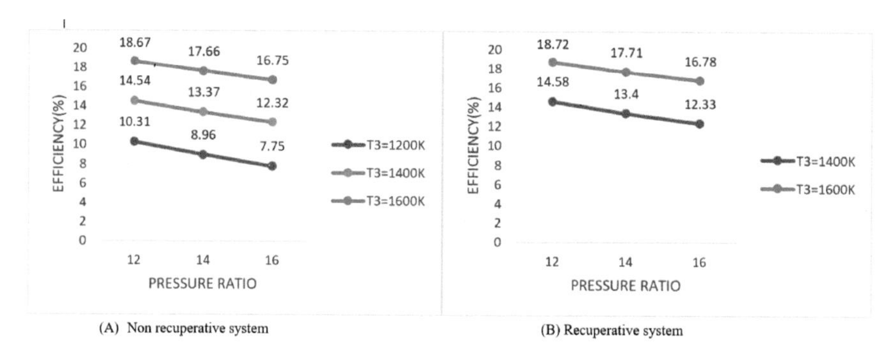

Figure 8.4 Variation in efficiency and pressure ratio for blend of 95% natural gas and 5% hydrogen.

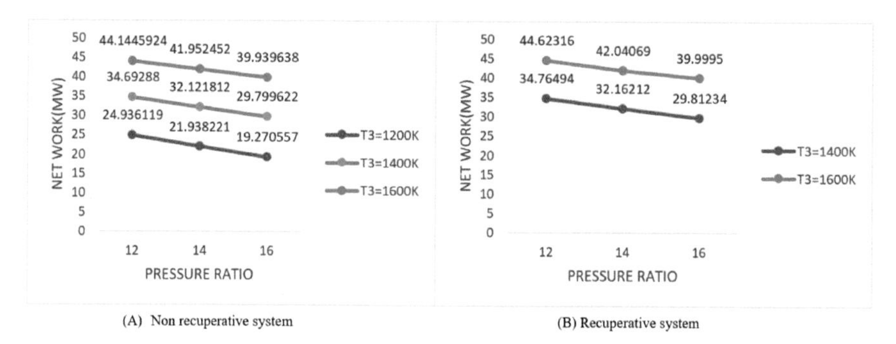

Figure 8.5 Variation of network and pressure ratio for blend of 90% natural gas and 10% hydrogen.

overall efficiency of the system since the amount of work produced is more and, at the same time, the amount of heat supplied to the system is less.

Scenario 2: Effect of 90% natural gas and 10% hydrogen fuel:

Figures 8.5 (a), 8.5 (b), 8.6 (a), and 8.6 (b) depict the variation in the net-work and efficiencies for alternating parameters for non-recuperative and recuperative systems, respectively, with changed fuel composition. The proportions of natural gas and hydrogen in the fuel are 90% and 10%, respectively. With the proportion of hydrogen increasing, the value of net-work and efficiency increases, and this can be attributed to hydrogen having three times the energy density of conventional natural gas, which affects the system output and efficiency in a positive manner. If the graphical results are examined, then it is observed that increasing the concentration of hydrogen

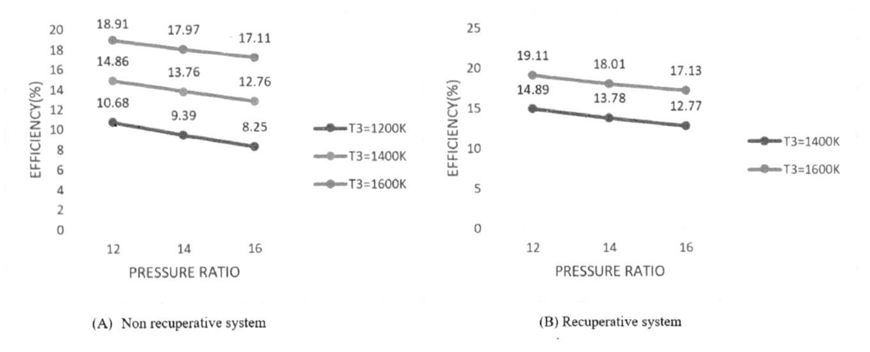

Figure 8.6 Variation in efficiency with pressure ratio for blend of 90% natural gas and 10% hydrogen.

contributes positively to the amount of work produced, and at pressure ratio 12, the average increase in the net-work is 6.33%. The regenerator effect is only presented at 1400 K and 1600 K turbine inlet temperatures because at low temperatures, the regenerative effect is negligible because of lower turbine outlet temperatures and hence can be neglected. Since the heat supplied to the system is less in the case of the regenerator and the value of the work produced is higher, the efficiency of the system naturally increases relative to the previous scenario where the hydrogen content was less.

Scenario 3: Effect of 85% natural gas and 15% hydrogen fuel:

Figures 8.7 (a), 8.7 (b), 8.8 (a), and 8.8 (b) depict the variation in the network and efficiencies for increasing turbine inlet temperature and pressure

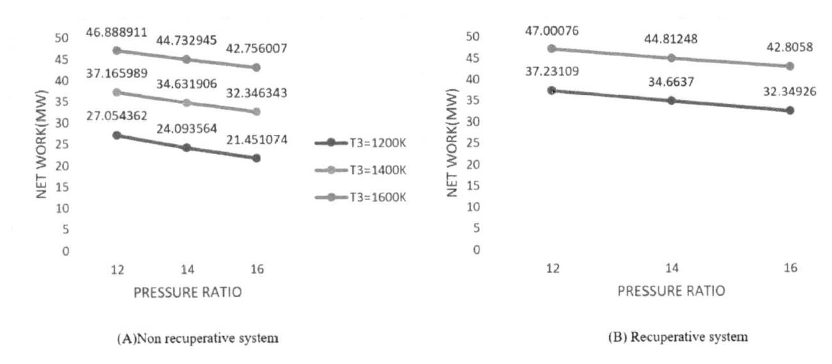

Figure 8.7 Variation in network with pressure ratio for blend of 85% natural gas and 15% hydrogen.

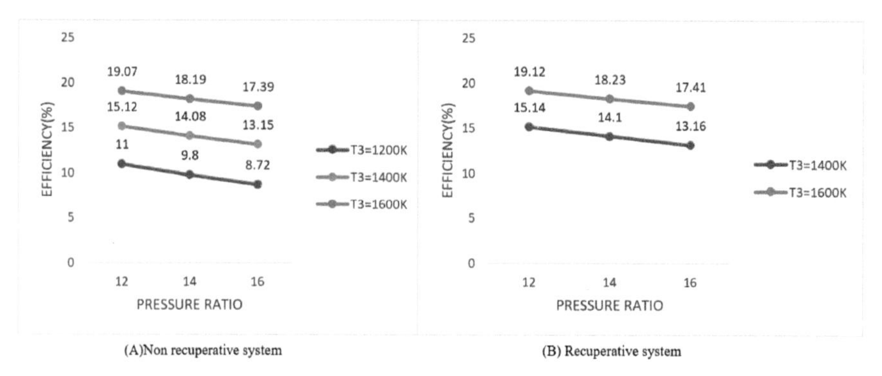

Figure 8.8 Variation in efficiency and pressure ratio for blend of 85% natural gas and 15% hydrogen.

ratio for non-recuperative and recuperative systems, respectively. The fuel composition has been changed in this scenario, and the proportions of natural gas and hydrogen in the fuel are 85% and 15%, respectively, and the increase in the hydrogen composition percentage increments the overall net-work and efficiency, and the trend of improving work output and efficiency with the increasing of the hydrogen percentage can be observed.

Scenario 4: Effect of 80% natural gas and 20% hydrogen fuel:

Figures 8.9 (a), 8.9 (b), 8.10 (a), and 8.10 (b) depict the variation in the net-work and efficiencies for increasing turbine inlet temperature and pressure ratio for non-recuperative and recuperative systems operating on 80% natural gas and 20% hydrogen. In this case, the amount of hydrogen

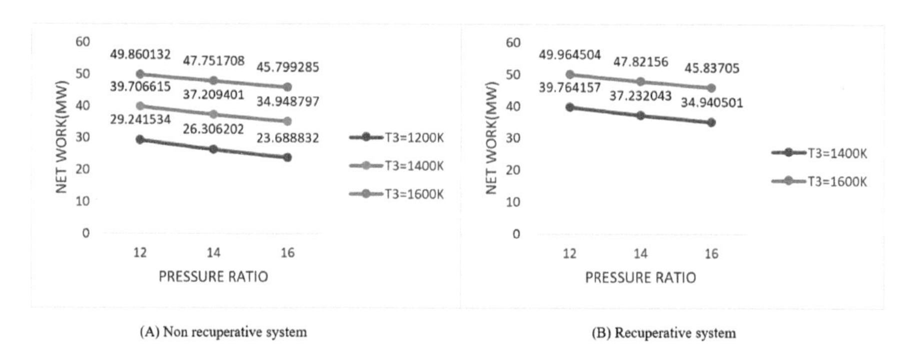

Figure 8.9 Variation in network with pressure ratio for blend of 80% natural gas and 20% hydrogen.

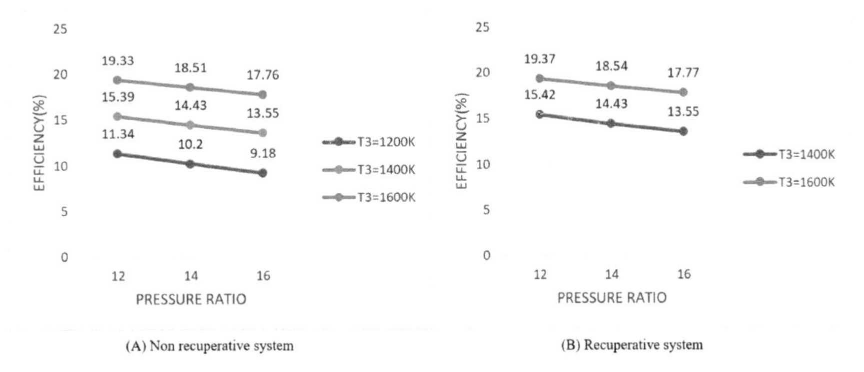

Figure 8.10 Variation in efficiency with pressure ratio for blend of 80% natural gas and 20% hydrogen.

used is the highest; so the amount of work done and the efficiency of the system are both at their highest levels. When compared to a fuel made of 95% natural gas and 5% hydrogen, the values of net-work and efficiency go up by a lot. The lower calorific value (LCV) of hydrogen is 119,986 kJ/mol, and this contributes tremendously to the overall work output when compared to natural gas, which has a lower calorific value of 50,000 KJ/mol. From the following scenarios, it can be concluded that the addition of hydrogen, even in a small percentage, will result in increased and improved work output and system efficiency.

Scenario 5: 95% natural gas, 2.5% hydrogen, and 2.5% steam effect:

The above scenario depicts the variation in net-work and efficiency of recuperative and non-recuperative gas turbine systems with the fuel as a blend of natural gas, hydrogen, and steam in varying proportions.

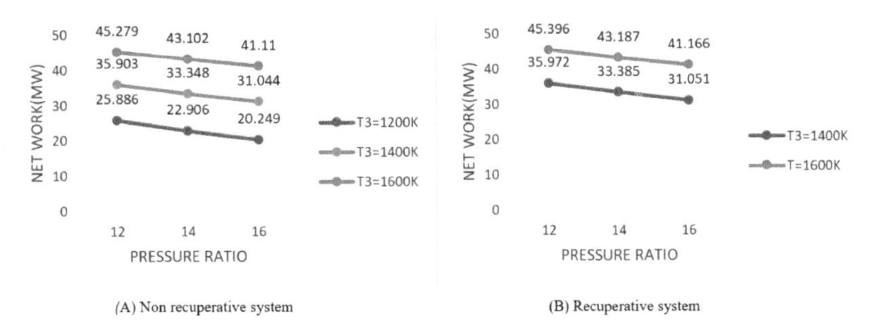

Figure 8.11 Variation in network with pressure ratio for blend 95% natural gas, 2.5% hydrogen, and 2.5% steam.

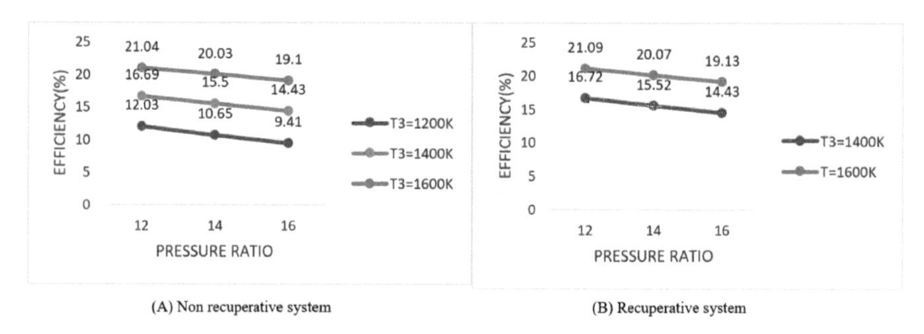

Figure 8.12 Variation in efficiency with pressure ratio for blend 95% natural gas, 2.5% hydrogen, and 2.5% steam.

Scenario 6: 90% natural gas, 5% hydrogen, and 5% steam effect:

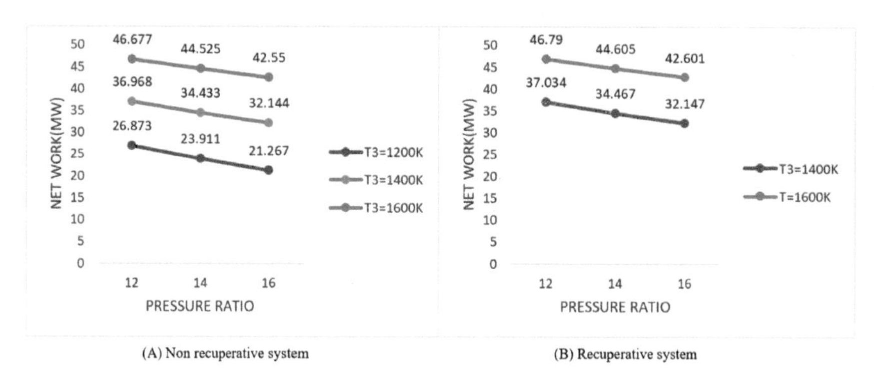

Figure 8.13 Variation in network with pressure ratio for blend 90% natural gas, 5% hydrogen, and 5% steam.

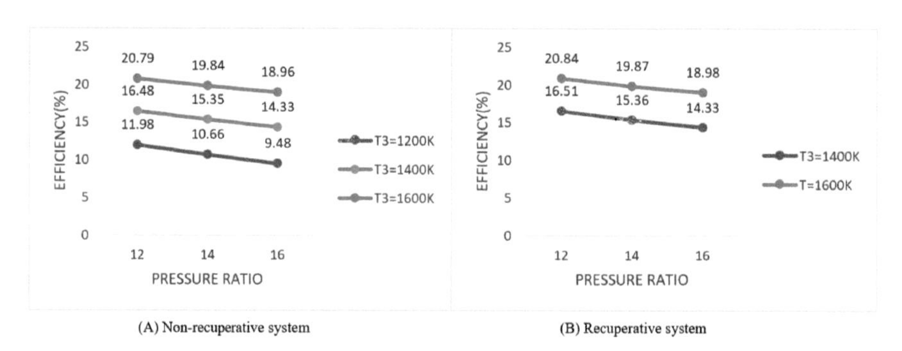

Figure 8.14 Variation in efficiency with pressure ratio for blend 90% natural gas, 5% hydrogen, and 5% steam.

Scenario 7: 85% natural gas, 7.5% hydrogen, and 7.5% steam effect:

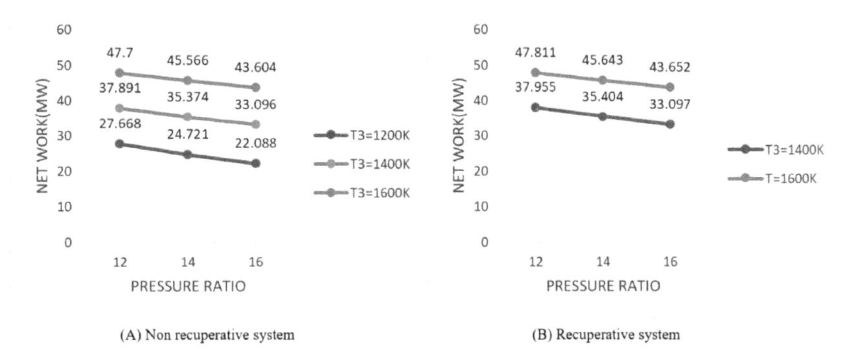

Figure 8.15 Variation in network with pressure ratio for blend 85% natural gas, 7.5% hydrogen, and 7.5% steam.

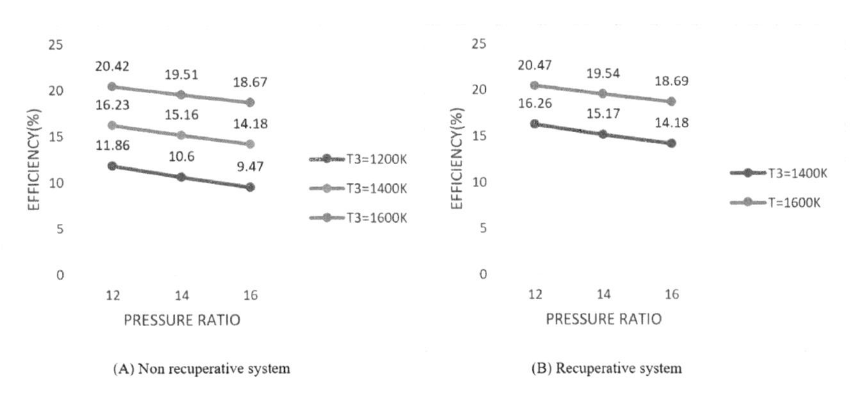

Figure 8.16 Variation in efficiency with pressure ratio for blend 85% natural gas, 7.5% hydrogen, and 7.5% steam.

Scenario 8: 80% natural gas, 10% hydrogen, and 10% steam effect:

Figures 8.11–8.18 depict the variations for the following variations in fuel proportions:

- 95% natural gas, 2.5% hydrogen, and 2.5% steam
- 90% natural gas, 5% hydrogen, and 5% steam
- 85% natural gas, 7.5% hydrogen, and 7.5% steam
- 80% natural gas, 10% hydrogen, and 10% steam

From the above figures, it can be observed that the addition of steam to the fuel positively impacts the values of net-work and system efficiency,

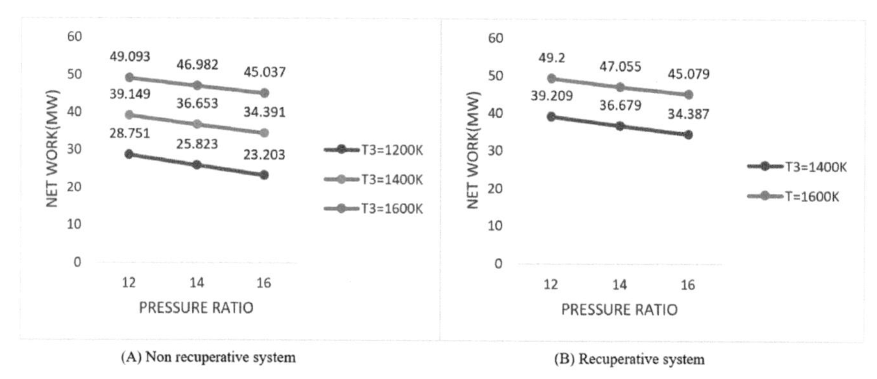

(A) Non recuperative system (B) Recuperative system

Figure 8.17 Variation in network with pressure ratio for blend 80% natural gas, 10% hydrogen, and 10% steam.

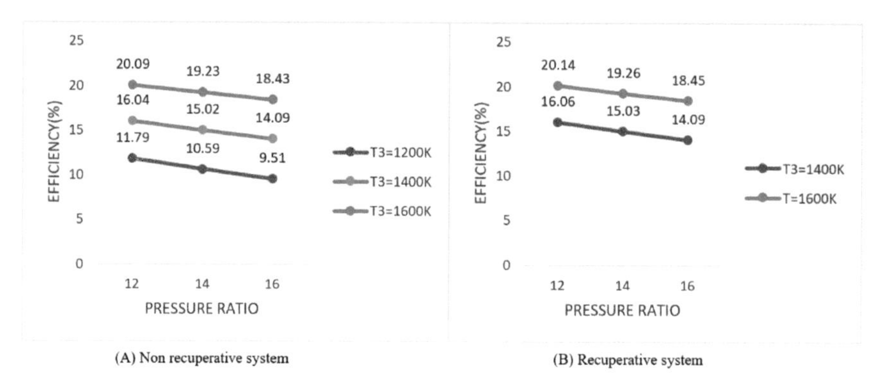

(A) Non recuperative system (B) Recuperative system

Figure 8.18 Variation in efficiency and pressure ratio for blend 80% natural gas, 10% hydrogen, and 10% steam.

and the value of net-work of the system at any particular pressure ratio and turbine inlet temperature is considerably higher than that of the system that is operating on natural gas and hydrogen fuel. This can be attributed to the fact that the addition of steam further complements the energy efficiency brought to the fuel by hydrogen, and this additional fuel helps the system achieve more work. But the efficiency of the system in this case decreases with an increasing percentage of steam. This can be explained by the following points. The efficiency of the gas turbine system is given by

$$\eta = \frac{W_{\mathrm{GT}} - W_{\mathrm{Com}}}{Q_{\mathrm{sys}}} \tag{8.24}$$

where Q_{sys} is the amount of heat supplied to the system and it is determined from the following formula:

$$Q_{\text{sys}} = m * \text{LCV}_{\text{fuel}}. \tag{8.25}$$

In the case of steam addition to the fuel, the overall heat of the system changes with increasing proportions of steam, as well as due to the fact that steam has a different lower calorific value at different temperatures and pressure ratios. This increases the amount of heat supplied to the system, thereby decreasing its efficiency. Therefore, it can be stated that addition of steam will positively benefit the system when it comes to the net-work output but at the cost of overall system efficiency.

8.5 Exergy Destruction and Efficiency

Figures 8.19–8.22 depict the exergy destruction and efficiencies of AC, CC, and GT for natural gas amalgamated with hydrogen and steam at varying proportions. The maximum exergy destruction is noticed in the combustion chamber. This phenomenon can be explained by the fact that fuel is injected and there is a vast difference in temperature. Exergy destruction affects the exergetic efficiencies of all the components involved, and the exergetic efficiencies will worsen with increasing exergy destruction. The exergetic efficiency of the combustion chamber can be noticed to be the lowest at

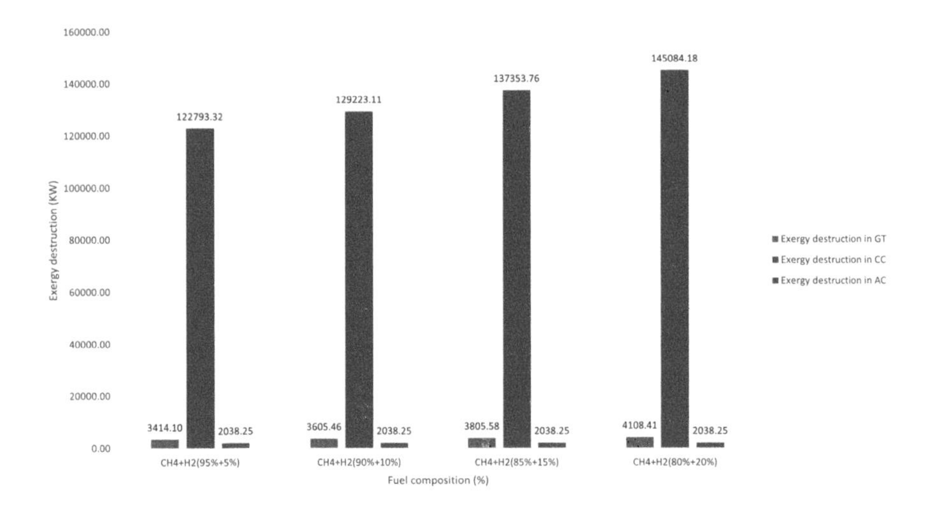

Figure 8.19 Variation in exergy destruction in all the components with fuel composition.

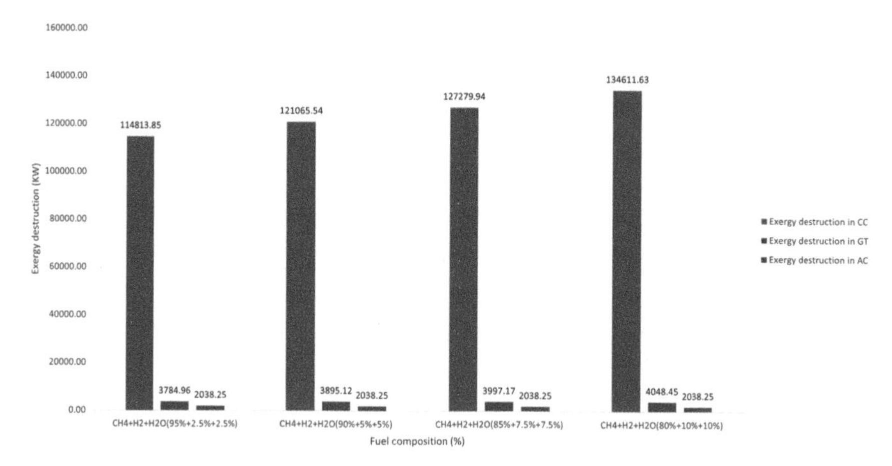

Figure 8.20 Variation in exergy destruction in all the components with fuel composition.

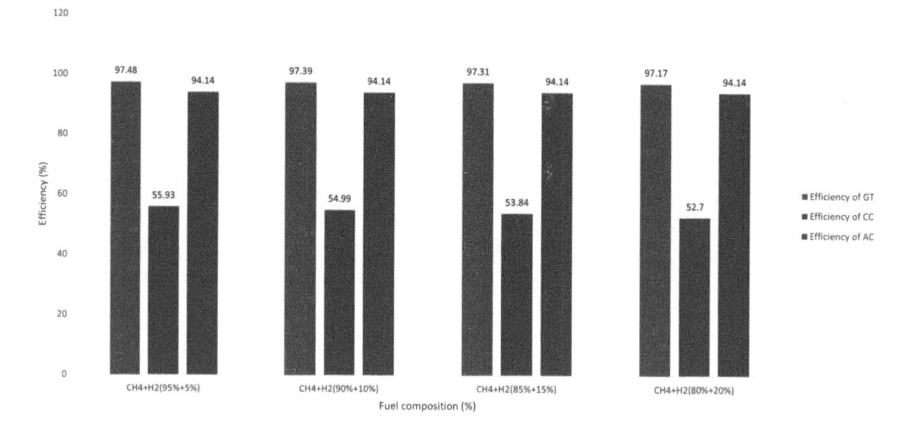

Figure 8.21 Exergetic efficiencies of the components with varying fuel composition.

approximately 55%–56%, and the exergetic efficiency of the air compressor is fixed at 94.14%. The figures are based on the physical exergies of different fuels and operating conditions, as physical exergy plays a major role in the overall exergy glow and destruction. The physical exergy consists of two main aspects, which are the exergy due to pressure and the exergy due to temperature, respectively. The exergy due to temperature changes with increasing inlet temperatures, but the main factor here is the exergy due to pressure. Since the calculations were done for varied pressure ratios, there were humongous amounts of values involved, and so for calculation purposes, a pressure ratio of 12 and an inlet temperature of 1200 K have been chosen.

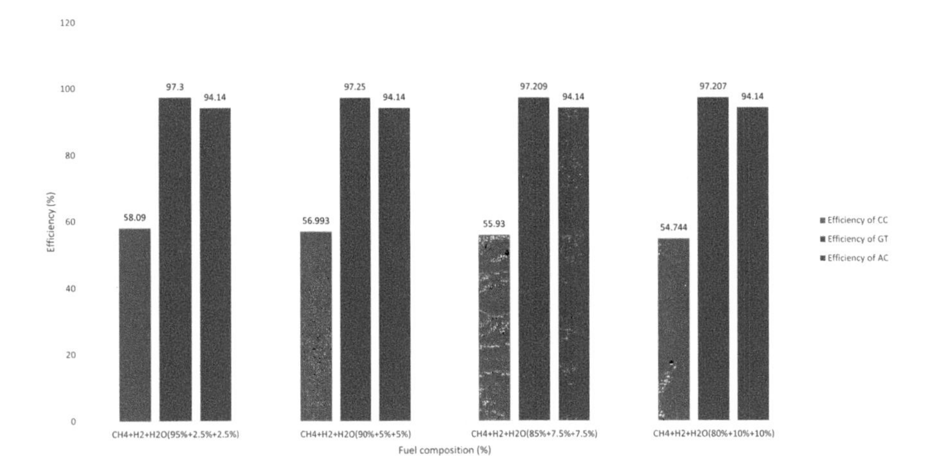

Figure 8.22 Exergetic efficiencies of the components with varying fuel composition

For this particular pressure ratio and inlet temperature, the value of exergy due to pressure varies due to the varying composition and proportions of the individual fuels. Exergy destruction has been calculated based on the values of physical exergy itself, and the exergy destruction is hugely dependent on the surrounding temperature. Incrementing the indicated parameter will exert influence on the exergy destruction, which will subsequently result in decrementing efficiency. Hence, to decrease the destruction and improve the efficiency of the individual components, the inlet temperature has to be kept as low as possible.

8.6 Conclusions

Hydrogen's momentousness in the power generation industry has been scrutinized thoroughly, and a major gap that needed addressing was the usage of hydrogen blended with other fuels for power generation purposes. In this chapter, an energy and exergy analysis of a gas turbine system working in different conditions with mixed fuels like natural gas, hydrogen, and steam has been done and analyzed. The following conclusions are drawn from this chapter:

- The analysis shows that adding hydrogen can be a very important part of improving the net-work output and the efficiency of the whole system better.

- Steam injection too helps in increasing the net-work output of the system while decreasing its efficiency, and this problem arises mainly due to the very high heat that is being supplied to the system.
- The system's exergy analysis shows that the combustion chamber experiences the greatest exergy destruction, and the significant temperature difference may be the cause of this phenomenon. Energy and exergy analysis is a very vital tool in understanding the potential limitations of the system, and it also provides an insight into how to deal with them.
- A novel method to increase the temperature at the inlet of the turbine can be figured out, which will ensure higher work output.
- As an alternative fuel, hydrogen can play a very important role in maximizing the efficiency of power generation and the total amount of power output. It is also good for the environment because it produces less pollution and has a smaller carbon footprint.
- Pure hydrogen-fueled power plants can achieve success in the aforementioned regards up to a certain threshold. So the blending of fuels to power the plants is a major game changer in this regard, as blending fuels will incorporate all the positive aspects of the fuels being blended, resulting in a higher power output with even greater efficiency while minimizing the cost of operation.

References

[1] Esclapez, L., Ma, P.C., Mayhew, E., Xu, R., Stouffer, S., Lee, T., Wang, H. and Ihme, M., 2017. Fuel effects on lean blow-out in a realistic gas turbine combustor. *Combustion and Flame, 181*, pp.82-99.

[2] Chiesa, P., Lozza, G. and Mazzocchi, L., 2005. Using hydrogen as gas turbine fuel. *J. Eng. Gas Turbines Power, 127*(1), pp.73-80.

[3] Aydin, K. and Kenanoğlu, R., 2018. Effects of hydrogenation of fossil fuels with hydrogen and hydroxy gas on performance and emissions of internal combustion engines. International journal of hydrogen energy, 43(30), pp.14047-14058.

[4] Kenanoğlu, R., Baltacıoğlu, M.K., Demir, M.H. and Özdemir, M.E., 2020. Performance & emission analysis of HHO enriched dual-fueled diesel engine with artificial neural network prediction approaches. International Journal of Hydrogen Energy, 45(49), pp.26357-26369.

[5] Baltacioglu, M.K., Kenanoglu, R. and Aydın, K., 2019. HHO enrichment of bio-diesohol fuel blends in a single cylinder diesel engine. International Journal of Hydrogen Energy, 44(34), pp.18993-19004.
[6] Arat, H.T., Baltacioglu, M.K., Tanç, B., Sürer, M.G. and Dincer, I., 2020. A perspective on hydrogen energy research, development and innovation activities in Turkey.
[7] Tanç, B., Arat, H.T., Conker, Ç., Baltacioğlu, E. and Aydin, K., 2020. Energy distribution analyses of an additional traction battery on hydrogen fuel cell hybrid electric vehicle. International Journal of Hydrogen Energy, 45(49), pp.26344-26356.
[8] Tanç, B., Arat, H.T., Baltacıoğlu, E. and Aydın, K., 2019. Overview of the next quarter century vision of hydrogen fuel cell electric vehicles. International Journal of Hydrogen Energy, 44(20), pp.10120-10128.
[9] Arat, H.T., Sürer, M.G., Gökpinar, S. and Aydin, K., 2020. Conceptual design analysis for a lightweight aircraft with a fuel cell hybrid propulsion system. Energy Sources, Part A: Recovery, Utilization, and Environmental Effects, pp.1-15.
[10] Bothien, M.R., Ciani, A., Wood, J.P. and Fruechtel, G., 2019, June. Sequential Combustion in Gas Turbines: The Key Technology for Burning High Hydrogen Contents With Low Emissions. In Turbo Expo: Power for Land, Sea, and Air (Vol. 58615, p. V04AT04A046). American Society of Mechanical Engineers.
[11] Gibrael, N. and Hassan, H., 2019. HYDROGEN-FIRED gas turbine for power generation with exhaust gas recirculation: emission and economic evaluation of pure hydrogen compare to natural gas.
[12] Božo, M.G., Vigueras-Zuniga, M.O., Buffi, M., Seljak, T. and Valera-Medina, A., 2019. Fuel rich ammonia-hydrogen injection for humidified gas turbines. Applied Energy, 251, p.113334.
[13] Tekin, N., Ashikaga, M., Horikawa, A. and Funke, H., 2018. Enhancement of fuel flexibility of industrial gas turbines by development of innovative hydrogen combustion systems. Gas Energy, 2, pp.1-6.
[14] Pavithran, A., Sharma, M. and Shukla, A.K., 2021. Oxy-fuel Combustion Power Cycles: A Sustainable Way to Reduce Carbon Dioxide Emission. Distributed Generation & Alternative Energy Journal, pp.335-362.

9

Fuel Cell and Hydrogen-based Hybrid Energy Conversion Technologies

Pranjal Kumar and Onkar Singh

Harcourt Butler Technical University, India
E-mail: sachanpranjal3@gmail.com, onkpar@rediffmail.com

Abstract

Over some time, the constantly increasing economic activities due to population growth have caused a rise in global electricity demand. This calls for enhancing electricity generation capacity. At the same time, the emerging concern for climate change due to polluting emissions getting added to the environment and the depletion of fossil fuels has necessitated the search for environmentally friendly direct energy conversion options. Hydrogen gas is the best alternative fuel to fossil fuels for energy conversion systems. Fuel cells in general and solid oxide fuel cells (SOFC) in particular have demonstrated good potential for direct electricity generation through the direct conversion of chemical energy into electricity. Consequently, rigorous efforts are being made by the scientific community to make SOFC-based electricity generation efficient. The energy loss minimization in the operation of SOFC is attempted by integrating different power cycles into it so that waste energy gets harnessed effectively. This chapter deals with an introduction to hydrogen production methods and fuel cell integrated systems, details of energy conversion technology in fuel cells and solid oxide fuel cells, waste energy assessment in SOFC, exploring possibilities of integrating power cycles to utilize the wastage of energy, and thermodynamic modeling of certain arrangements for a better understanding of how SOFC-integrated cycles can offer better energy efficiency. The synergetic operation of SOFC with different power cycles operating on various working fluids holds the future for the availability of reasonably clean energy.

Keywords: SOFC, GT, combined cycle, modeling

9.1 Introduction

The escalating pollution, electricity demand, and depletion, as well as the limited amount of fossil fuels, are the major concerns for society. Getting energy from renewable sources like the sun, geothermal, and wind energy is also in its emerging phase. But the non-availability of a continuous supply of energy makes them less efficient technologies. Therefore, the application of hydrogen gas as a fuel in the energy conversion systems is an alternative source of clean energy irrespective of the energy production from fossil fuels and its renewability, along with zero emission of dangerous pollutants. But the storage of hydrogen gas is more difficult compared to other gases; therefore, continuous production and utilization are the alternatives for which research is going on. Because hydrogen has a lower volume power density, it requires an extensive storage tank with greater pressure than other gaseous fuels (Lowe-Wincentsen, 2013). As a result, developing new storage technologies to maximize energy density per volume is critical.

Hydrogen's small weight makes it a solitary and innovative energy carrier with a wide range of applications. Water vapor is produced when hydrogen gas combustion takes place, and its 141.9 kJ/g calorific value is 3 times that of 47 kJ/g of gasoline and 2.6 times that of 54 kJ/g of natural gas (Nanda et al., 2017). There are different technologies through which hydrogen gas can be generated, as shown in Figure 9.1. Hydrogen gas is generally produced through the steam reforming of natural gas. However, according to Acar and Dincer (2014), producing hydrogen gas from fossil fuels is one of the most environmentally damaging procedures. As a result, large-scale, cost-effective, renewable, and environmentally friendly hydrogen generation techniques are urgently needed.

Electrochemical water dissociation is the splitting of water into hydrogen and oxygen. This is done by sending electricity through electrochemical cells. In a typical electrolysis machine, the negatively charged anode and positively charged cathode electrodes are submerged in an electrolyte. Hydrogen gas is released in the cathode after an electric current passes through it, whereas oxygen gas collects near the anode (Levene et al., 2007). One of the best emerging techniques for renewable hydrogen gas production is

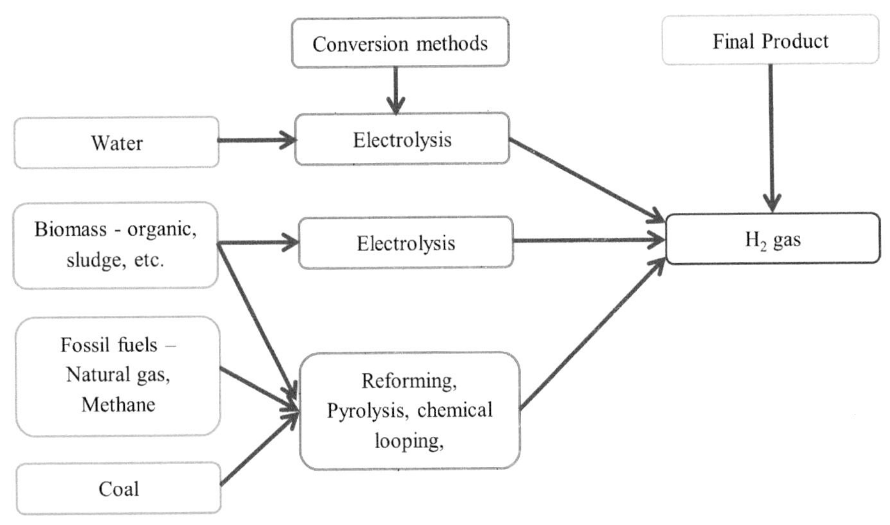

Figure 9.1 Schematic of hydrogen gas generation.

the redox reaction, which occurs when two half-reactions occur at the same time. An oxidation half-reaction known as the oxygen evolution reaction (OER) happens at the electrochemical cell's anode. At the electrochemical cell's cathode, a reduction half-reaction, also known as the hydrogen evolution reaction (HER), occurs. Eqn (9.1) and (9.2) represent OER and HER simultaneously.

$$2H_2O \rightarrow 4H^+ + O_2 + 4e^- \tag{9.1}$$

$$2H^+ + 2e^- \rightarrow H_2 \tag{9.2}$$

Fuel cell technology transforms chemical energy directly into electrical energy with less emissions or zero hazardous pollutants and is not constrained by the Carnot cycle (Haseli, 2018). Moreover, the fuel cell can create electricity by converting the chemical energy directly. Different types of fuel cells are present in the world market. A solid oxide fuel cell (SOFC) is one that operates at a high temperature. Because of the high operating temperature of SOFC, different fuels could also be supplied for its electricity generation. Additionally, SOFC generates the exhaust at a high temperature. This waste heat can be recovered by integrating other energy conversion technologies, like a gas turbine, steam turbine, organic Rankine cycle, etc., with the SOFC. The energy rejected at the exit of the SOFC can be employed for different useful activities like hot water production, cooling effects, space heating, etc.

The combined technology power plants are highly efficient in comparison to the solitary technology plants.

In this chapter, fuel cell and hydrogen-based hybrid energy conversion technologies have been presented. The thermodynamic analyses of two systems are presented: an SOFC and a gas turbine system. Some details of the parametric studies to assess the impact of operating-condition changes on the system's performance are also presented herein.

9.2 Types of Fuel Cells

There are different kinds of fuel cells based on the temperature range in which they work and how they are fueled (Kumar and Singh, 2022). For example, alkaline fuel cells operate at 90–100 °C and generate hydrogen gas (Sommer et al., 2016); proton exchange membrane fuel cells operate at 50–100 °C and generate hydrogen gas (Yu et al., 2022); molten carbonate fuel cells operate at 600–700 °C and generate carbon monoxide gas (Souleymane et al., 2022); and phosphoric acid fuel cells operate at 150–200 °C and generate hydrogen. Among all of them, proton exchange membrane fuel cells (PMEFC) and SOFC are more important than others.

PMEFC runs at a low temperature. It is used for light-duty vehicles. On the other hand, an SOFC produces electricity while operating at a high temperature. They are useful for stationary applications like power plants. The high operating temperature of the SOFC releases the high-temperature exhaust, which enables the opportunity for integration with other energy conversion technologies like a gas turbine, steam turbine, internal combustion engine, etc.

9.3 Fuel Cell Energy Conversion Technologies

A fuel cell transforms the chemical energy of the fuel into electricity without any intermittent conversion of energy. Only water and waste heat were released as the fuel cell's output. Reduction in the energy conversion step leads to improvement in fuel cell efficiency. Therefore, fuel cells become highly efficient. The basic structure of the fuel cell has an electrode and electrolyte. In the fuel cell, hydrogen gas and air are fed into the anode and cathode sides simultaneously. Then the hydrogen releases the electrons in the anode, which move to the cathode side via an external circuit and produce electricity. After receiving the electron at the cathode side, the oxygen ion moves toward the anode side and reacts with the hydrogen to create water and electricity.

9.4 Hydrogen Generation with Fuel Cells

At the end of the fuel cell, the temperature is very high, which could be used for a certain purpose. The proton exchange membrane electrolyzer (PEME) or alkaline water electrolysis (AWE) can both be used to make hydrogen gas. Alkaline water electrolysis is the most common and financially viable way to separate water. It is strong and can make many megawatts of hydrogen. Straightforward design and a low-cost electrolyte are two of its strengths (Turner et al., 2008). Alkaline electrolysis, on the other hand, has limitations such as less than 0.6 A/cm^2 of current densities (Buttlerand and Spliethoff, 2018), which results in a non-complex design, and significant gas transitions between the anode and cathode (Carmo et al., 2013). In contrast to these drawbacks, PEM water electrolyzers have been designed that transport hydrogen ions over a polymer membrane rather than a liquid electrolyte. PEM water electrolyzers can reach current densities of up to 10.0 A/cm^2 (Villagra and Millet, 2019), but they are usually only employed up to 2.0 A/cm^2 (Buttlerand and Spliethoff, 2018). In comparison to alkaline electrolysis, this, along with the thin membrane utilization, permits for a relatively compact design and much less gas crossover. PEMEs are the best-suited technology to contribute to this stabilization in association with PEM fuel cells, amidst various potential alternatives, for example, redox flow and Li-ion batteries. Hydrogen can be created and stored in the event of excess electricity generation. When electricity generation becomes insufficient in the future, the previously created hydrogen could be turned into electricity through hydrogen fuel cells. As a result, the grid is stabilized, and apexes and troughs in renewable output are reduced (Allidières et al., 2019).

PEME has a liquid gas diffusion layer (LGDL) at the anode and a gas diffusion layer (GDL) at the cathode. It also has a catalyst-coated membrane, end plates, and gaskets to stop leaks. Water enters the PEME through the anode-side end-plate and is carried across the active region via the flow channels. Field water is liquid within the temperature range in which PEMEs are generally used. Water then passes through the porous LGDL on its way to the anode catalyst layer (CL), where it is oxidized into protons, electrons, and oxygen. The protons are carried through the membrane to the cathode CL, where they are converted to hydrogen. However, oxygen must be delivered from the anode CL to the flow channels, where it is carried out of the PEME cell with the unreacted water via the LGDL (Maier et al., 2022). For space heating, water heating, and gas cooking, hydrogen can be used instead of natural gas. Out of all the engineering considerations, the Wobbe index is the most straightforward and widely used comparison statistic for assessing how well appliances work with various types of gases (Dodds et al., 2014).

9.5 Requirement and Development of SOFC Integrated Cycle

SOFCs have doped zirconia solid oxide electrolytes. Due to an operating temperature in the range of 1073–1273 K, the electrolyte easily conducts ions and transfers them from one side to another of the fuel cell. The exhaust temperature of the SOFC is high, nearly 753–1123 K. Therefore, there is an opportunity to integrate the fuel cell with other technologies in parallel or series. Generally, other technologies are utilized with the SOFC as a bottoming cycle for achieving high efficiency while augmenting power. The electrolyte used in the SOFC is solid; therefore, different types of fuel, like natural gas, biogas, etc., could be supplied in the SOFC for electrical energy. Generally, internal reforming is less recommended in comparison to external reforming because the chance of carbon deposition increases at the electrode. However, reforming of the fuel other than hydrogen is done internally at the SOFC because of the high operating temperature. A single cell of the fuel cell has low voltage; therefore, the number of fuel cells is arranged together in the stack with the help of the intermittent.

A group of researchers is always looking into fuel cells and other hybrid systems that change energy to find out which one is the best for making power and having cooling effects. Singh et al. (2022) looked into an integrated system with SOFC, an absorption cycle, a waste heat recovery system, and hydrogen storage. They have done a parametric analysis of the combined system to see how the key parameters, like current density, fuel cell inlet temperature, etc., affect the system as a whole. Wang et al. (2022) used the first and second laws of thermodynamics to look at the SOFC integrated system. This system can make hydrogen and compress and store carbon dioxide. Beigzadeh et al. (2021) investigated several fuels and used the SOFC-GT system as a framework. The natural gas-fueled system was shown to have superior electrical and energy efficiency, reaching 72.71% and 61.29%, respectively. Chitgar and Emadi (2021) presented a local power generation SOFC-GT model capable of producing both hydrogen and freshwater.

Under ideal circumstances, the system's exergy efficiency was 59.4% and the overall price per unit was 23.6 $/GJ, according to the thermodynamic and economic assessments. In a multi-generation cycle, Emadi et al. (2020) explored the idea of coupling SOFC, GT, and a dual-loop ORC. Liquefied natural gas (LNG) was used as a heat sink to recuperate the LNG stream's cold energy. The ideal working fluid combination for dual-loop ORC, according to the optimization results, is R601-Ethane. Furthermore, a thermos-economic study revealed that, when compared to SOFC-GT and plain SOFC systems, the suggested system may achieve a lower levelized cost of power by 12.9% and 73.9%, respectively.

Habibollahzade et al. (2019) compared the impacts of several gasification agents in a biomass-gasified SOFC system, and a solid oxide electrolyzer that uses energy generated by a gas turbine. Gasifying agent carbon dioxide results in lower carbon dioxide emissions and more hydrogen production, as seen in the results. Ghaffarpour et al. (2018) investigated the thermodynamics and economics of an interconnected system that included an SOFC, a GT, and a Rankine cycle with biomass as a feedstock. It was discovered that the pine sawdust fuel has stronger economic and thermodynamic results than the gas emitted from solid urban wastes and poultry byproducts after studying thermodynamic parameters such as the operating pressure and temperature of the system, fuel flow, current density, and compression ratio in the compressor on the performance of the system and the total expenses of the hybrid system. Zhuang et al. (2017) studied a novel integration of the hydrogen fuel cell and carbon fuel cell with the natural gas fed-up system and also presented carbon sequestration.

9.6 Modeling and Analysis of Solid Oxide Fuel Cell Integrated Cycle

The hydrogen-fed fuel cell system involves an electrochemical reaction to produce free electrons to deliver electricity and water as an exhaust product. The overall reaction involved is as follows:

$$H_2+O_2\rightarrow 0.5H_2O. \tag{9.3}$$

The voltage of the fuel cell (Hosseinpour et al., 2017) can be evaluated as follows:

$$V = V_{\text{Nerst}} - V_{\text{loss}} \tag{9.4}$$

where V, V_{Nerst}, and V_{loss} represent the voltage of the fuel cell, the Nerst potential, and the voltage loss that occurs in the fuel cell in the sum of ohmic voltage loss (V_{ohmic}), concentration voltage loss (V_{conc}), and activation voltage loss (V_{act}).

Nerst voltage (Oryshchyn et al., 2018) can be calculated as follows:

$$V_{\text{Nerst}} = 1.253 - 2.4516 * 10^{-4} * T_{\text{SOFC}} - \left(\frac{R.T_{\text{SOFC}}}{4F}\right) * \ln\left(\frac{p_{\text{H}_2\text{O}}^2}{p_{\text{H}_2}^2 \cdot p_{\text{O}_2}}\right) \tag{9.5}$$

where T_{SOFC}, R, F, and P_i denote the fuel cell temperature, the universal gas constant (8.314 J/molK), the Faraday constant (96,485 C/mol), and the partial pressure of products and reactants of the electrochemical reaction. The occurring voltage losses in the fuel cell are calculated by using eqn (9.4)–(9.8) (Chitsaz et al., 2015; Ranjbar et al., 2014).

$$V_{\text{ohmic}} = (R_c + \rho_c L_c + \rho_a L_a + \rho_e L_e + \rho_{\text{int}} L_{\text{int}}) \,.\, J \tag{9.6}$$

where R_c, ρ_c, ρ_a, ρ_e, ρ_{int}, L_c, L_a, L_e, L_{int}, and J represent the resistivity contact, the electrical resistivity of the cathode component of the cell, the electrical resistivity of the anode component of the cell, the electrical resistivity of the electrode of the cell, the electrical resistivity of the intermittent cell, the thickness of cathode, the thickness of anode, the thickness of electrolyte, thickness of intermittent, and current density, respectively.

$$\rho_e = \left(3.34 * 10^4 * e^{(-103000/T_{\text{SOFC}})}\right)^{-1} \tag{9.7}$$

$$\rho_a = \left(\frac{95 * 10^6}{T_{\text{SOFC}}} * e^{(-1150/T_{\text{SOFC}})}\right)^{-1} \tag{9.8}$$

$$\rho_c = \left(\frac{42 * 10^6}{T_{\text{SOFC}}} * e^{(-1200/T_{\text{SOFC}})}\right)^{-1} \tag{9.9}$$

$$\rho_{\text{int}} = \left(\frac{9.3 * 10^6}{T_{\text{SOFC}}} * e^{(-1100/T_{\text{SOFC}})}\right)^{-1}. \tag{9.10}$$

Ohmic voltage loss can be enumerated by dealing with the above equation while taking fuel cell temperature, cell component thickness, and current density as input parameters.

Another voltage loss named as concentration voltage loss is the summation of concentration voltage loss at anode ($V_{conc,a}$) and at the cathode ($V_{conc,c}$). Mathematical expression for determining the concentration voltage loss is depicted in eqn (9.11)– (9.15):

$$V_{\text{conc}} = V_{\text{conc},a} + V_{\text{conc},c} \tag{9.11}$$

$$V_{\text{conc},a} = \left(\frac{R.T_{\text{SOFC}}}{2F}\left[\ln\left(1 + \frac{P_{H_2}.\,J}{P_{H_2O}.\,J_{as}}\right) - \ln\left(1 - \frac{J}{J_{as}}\right)\right]\right) \tag{9.12}$$

$$V_{\text{conc},c} = -\left(\frac{R.T_{\text{SOFC}}}{4F}.\ln\left(1 - \frac{J}{J_{cs}}\right)\right) \tag{9.13}$$

$$J_{as} = \frac{2.F.P_{\text{H}_2}.D_{\text{aeff}}}{R.T_{\text{SOFC}}.L_a} \tag{9.14}$$

$$J_{cs} = \frac{4.F.P_{\text{O}_2}.D_{\text{ceff}}}{\left(\left(\frac{P_{\text{exit, c}} - P_{\text{O}_2,c}}{P_{\text{exit, c}}}\right) R.T_{\text{SOFC}}.L_c\right)} \tag{9.15}$$

where J_{as}, J_{cs}, D_{ceff}, and D_{aeff} represent the exchange current at the anode (J_{as}) and cathode (J_{cs}), and effective gaseous diffusivity through the cathode (D_{ceff}) and the anode (D_{aeff}).

Activation voltage loss (V_{act}) can be determined by eqn (9.16):

$$V_{\text{act}} = V_{\text{act},a} + V_{\text{act},c} \tag{9.16}$$

$$V_{\text{act},a} = \left(\frac{R.T_{\text{SOFC}}}{F}.\sinh^{-1}\left(\frac{J}{2.J_{oa}}\right)\right) \tag{9.17}$$

$$V_{\text{act},c} = \left(\frac{R.T_{\text{SOFC}}}{F}.\sinh^{-1}\left(\frac{J}{2.J_{oc}}\right)\right). \tag{9.18}$$

Power output from the single fuel cell ($P_{fuelcell}$) can be enumerated as follows:

$$P_{\text{fuel cell}} = \eta_{\text{DC/AC}}.\ J.A.V. \tag{9.19}$$

The variation of the voltage with the current density is shown in Figure 9.2. Generally, fuel cells are used in stack form. Therefore, they have a number of fuel cells. Now, obtaining the power generated by the fuel cell stack is equal to the multiplication of the fuel cell number and the power generated by the single fuel cell.

Here, $\eta_{\text{DC/AC}}$is the efficiency of conversion from DC to AC. Therefore, the fuel cell efficiency can be determined by eqn (9.20):

$$\eta_{\text{DC/AC}} = \frac{P_{\text{fuel cell}}}{\phi.\text{LHV}} \tag{9.20}$$

where ϕ is the molar flow rate of the fuel in the fuel cell. This molar flow rate can be converted into the mass flow rate.

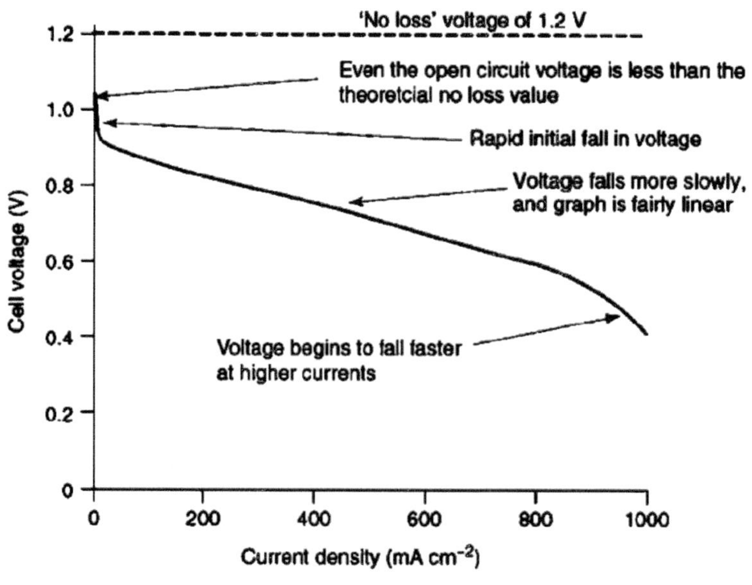

Figure 9.2 Representation of the voltage variation with the current density (Larminie and Dicks, 2003).

The detailed modeling of the gas turbine cycle based on the first and second laws of thermodynamics is presented by Kumar and Singh (2019). The study detailed a thermo-economic analysis of the SOFC-GT combined arrangement. Mass balance and energy balance, as shown in eqn (9.21) and (9.22), have been done for enumerating the mass flow rate, temperature, and enthalpy at the point of the system.

$$\sum m_{\text{inlet}} = \sum m_{\text{exit}} \tag{9.21}$$

$$\sum m_{\text{inlet}} h_{\text{inlet}} + Q = \sum m_{\text{exit}} h_{\text{exit}} + W \tag{9.22}$$

where W and Q denote the work generated or consumed by the system and heat input.

9.7 Fuel Cell and Hydrogen-based Hybrid Energy Conversion

Kumar and Singh (2019) say that hydrogen fuel cell systems convert energy very well, with 50%–60% efficiency when used alone. Due to its high energy conversion efficiency and ability to work with other energy conversion systems, the fuel cell is a good way to meet the growing need for energy. Researchers have been looking at different ways to combine fuel cells with other energy conversion systems for the past 10 years. For example, a biomass integrated SOFC system has been investigated by Emadi et al. (2020) to improve the efficiency of the power plant. SOFC-GT hybrid systems (Shukla et al., 2018; Zhang et al., 2022), SOFC-ORC systems (Liu et al., 2019), SOFC and internal combustion engines (Kim et al., 2020), etc., are other hybrid systems to meet energy demand.

9.8 Conclusions

Hydrogen gas utilization as fuel is a viable option due to its capability of producing clean energy and environmentally friendly water vapor as a byproduct. Hydrogen gas-fed hybrid energy conversion technologies are the most important technology for power production to meet the energy demand with the same amount of fuel utilization in the system. A fuel cell is an interesting arrangement for energy conversion, as its performance is not restricted by the limitations imposed by the Carnot cycle. When the hydrogen gas is directly fed into the fuel cell, it is highly efficient in comparison to other fuel-fed systems.

References

[1] Acar C, Dincer I, Comparative assessment of hydrogen production methods from renewable and non-renewable sources. International journal of hydrogen energy 2014;39(1):1-2.

[2] Allidières L, Brisse A, Millet P, Valentin S, Zeller M, On the ability of pem water electrolysers to provide power grid services. International Journal of Hydrogen Energy 2019;44(20):9690-700.

[3] Beigzadeh Milad, Pourfayaz Fathollah, Ghazvini Mahyar, Ahmadi Mohammad H., Energy and exergy analyses of solid oxide fuel cell-gas turbine hybrid systems fed by different renewable biofuels: A comparative study. Journal of Cleaner Production 2021;280:124383.

[4] Buttler A, Spliethoff H, Current status of water electrolysis for energy storage, grid balancing and sector coupling via power-to-gas and power-to-liquids: A review. Renewable and Sustainable Energy Reviews 2018;82:2440-54.

[5] Carmo M, Fritz DL, Mergel J, Stolten D, A comprehensive review on PEM water electrolysis. International journal of hydrogen energy 2013;38(12):4901-34.

[6] Chitgar N., Emadi M. A., Development and exergoeconomic evaluation of a SOFC-GT driven multi-generation system to supply residential demands: Electricity, fresh water and hydrogen. International of Hydrogen Energy 2021;46(34):17932-17954.

[7] Chitsaz A., Mahmoudi S. M. S., Rosen M. A., Greenhouse gas emission and exergy analyses of an integrated trigeneration system driven by a solid oxide fuel cell. Applied Thermal Engineering 2015;86:81–90.

[8] Dodds PE, Staffell I, Hawkes AD, Li F, Grünewald P, McDowall W, Ekins P. Hydrogen and fuel cell technologies for heating: A review. International Journal of Hydrogen Energy 2015;40(5):2065-83.

[9] Emadi M A, Karimi M H, Chitgar N, Ahmadi P, Rosen M A. Performance assessment and optimization of a biomass-based solid oxide fuel cell and micro gas turbine system integrated with an organic Rankine cycle. International Journal of Hydrogen Energy 2020;45(11):6262-77.

[10] Emadi Mohammad Ali, Chitgar Nazanin, Oyewunmi Oyeniyi A., Markides Christos N., Working-fluid selection and thermoeconomic optimisation of a combined cycle cogeneration dual-loop organic Rankine cycle (ORC) system for solid oxide fuel cell (SOFC) waste-heat recovery. Applied Energy 2020;261:114384.

[11] Ghaffarpour Z., Mahmoudi M., Mosaffa A. H., Farshi L. Garousi, Thermoeconomic assessment of a novel integrated biomass based power generation system including gas turbine cycle, solid oxide fuel cell and Rankine cycle. Energy Conversion and Management 2018;161:1-12.

[12] Habibollahzade Ali, Gholamian Ehsan, Behzadi Amirmohammad, Multi-objective optimization and comparative performance analysis of hybrid biomass-based solid oxide fuel cell/solid oxide electrolyzer cell/gas turbine using different gasification agents. Applied Energy 2019;233-234:985-1002.

[13] Haseli Y, Maximum conversion efficiency of hydrogen fuel cells. International Journal of Hydrogen Energy 2018;43:9015–9021.

[14] Hosseinpour J., Sadeghi M., Chitsaz A., Ranjbar F., Rosen M. A., Exergy assessment and optimization of a cogeneration system based on a solid oxide fuel cell integrated with a Stirling engine, Energy Conversion and Management 2017;143:448–458.

[15] Kim Y S, Lee Y D, Ahn K Y, System integration and proof-of-concept test results of SOFC–engine hybrid power generation system. Applied Energy 2020;277:115542.
[16] Kumar P, Singh O., A review of solid oxide fuel cell based hybrid cycles. International Journal of Energy Research 2022;46(7):8560-8589.
[17] Kumar P., Singh O., "Thermodynamic Analysis of Solid Oxide Fuel Cell-Gas Turbine-Organic Rankine Cycle Combined System". International Journal of Material Science and Mechanical Engineering (JMSME) 2019;6(2):98-102
[18] Kumar P., Singh O., Exergo-Economic Study of SOFC-Intercooled and Reheat type GT-VARS-ORC Combined Power and Cooling System. Journal of the Institution of Engineers (India): Series C 2021;102(5):1153-1166.
[19] Kumar P., Singh O., Thermoeconomic analysis of SOFC-GT-VARS-ORC combined power and cooling system. International Journal of Hydrogen Energy 2019;44:27575-27586.
[20] Larminie J, Dicks A, Fuel cell systems explained 2nd ed. Wiley 2003.
[21] Levene J I, Mann M K, Margolis R M, Milbrandt A, An analysis of hydrogen production from renewable electricity sources. Solar energy 2007;81(6):773-80.
[22] Liu Y, Han J, You H. Performance analysis of a CCHP system based on SOFC/GT/CO2 cycle and ORC with LNG cold energy utilization. International Journal of Hydrogen Energy 2019;44(56):29700-10.
[23] Lowe-Wincentsen D, Alternative Fuels Data Center: Reference Reviews 2013.
[24] Maier M, Smith K, Dodwell J, Hinds G, Shearing PR, Brett DJ. Mass transport in PEM water electrolysers: A review. International Journal of Hydrogen Energy 2022;47(1):30-56.
[25] Nanda S, Li K, Abatzoglou N, Dalai AK, Kozinski J A, Advancements and confinements in hydrogen production technologies. Bioenergy systems for the future 2017;373-418. Woodhead Publishing.
[26] Oryshchyn D., Harun N. F., Tucker D., Bryden K. M., Shadle L., Fuel utilization effects on system efficiency in solid oxide fuel cell gas turbine hybrid systems. Applied Energy 2018;228:1953-1965.
[27] Ranjbar F., Chitsaz A., Mahmoudi S.M.S., Khalilarya S., Rosen M.A., Energy and exergy assessments of a novel trigeneration system based on a solid oxide fuel cell, Energy Conversion and Management 2014;87:318–327.

[28] Singh U. R., Kaushik A. S., Bhogilla S. S., A novel renewable energy storage system based on reversible SOFC, hydrogen storage, Rankine cycle and absorption refrigeration system. Sustainable Energy Technologies and Assessments 2022;51:101978.
[29] Shukla, A. K., Gupta, S., Singh, S. P., Sharma, M., & Nandan, G. (2018). Thermodynamic performance evaluation of SOFC based simple gas turbine cycle. International Journal of Applied Engineering Research, 13(10), 7772-7778.
[30] Sommer E. M., Vargas J. V. C., Martins L. S., Ordonez J. C., The maximization of an alkaline membrane fuel cell (AMFC) net power output. International Journal of Energy Research 2016;40:924-939.
[31] Souleymane C, Zhao J, Li W. Efficient utilization of waste heat from molten carbonate fuel cell in parabolic trough power plant for electricity and hydrogen coproduction. International Journal of Hydrogen Energy 2022;47(1):81-91.
[32] Turner J, Sverdrup G, Mann M K, Maness P C, Kroposki B, Ghirardi M, Evans R J, Blake D, Renewable hydrogen production. International journal of energy research 2008;32(5):379-407.
[33] Villagra A, Millet P, An analysis of PEM water electrolysis cells operating at elevated current densities. International Journal of Hydrogen Energy 2019;44(20):9708-17.
[34] Wang H., Zhao H., Du H., Zhao Z., Zhang T., Thermodynamic performance study of a new diesel-fueled CLHG/SOFC/STIG cogeneration system with CO2 recovery. Energy 2022:123326.
[35] Wilailak S., Yang J. H., Heo C. G., Kim K. S., Bang S. K., Seo I. H., Zahid U., Lee C. J, Thermo-economic analysis of phosphoric acid fuel-cell (PAFC) integrated with organic ranking cycle (ORC). Energy 2021;220:119744.
[36] Xu Haoran, Chen Bin, Tan Peng, Cai Weizi, He Wei, Farrusseng David, Ni Meng, Modeling of all porous solid oxide fuel cells. Applied Energy 2018;219:105-13.
[37] Yaqoob L., Noor T., Iqbal N., Recent progress in development of efficient electrocatalyst for methanol oxidation reaction in direct methanol fuel cell. International Journal of Energy Research 2021;45(5):6550-83.
[38] Yu Z., Feng C., Lai Y., Xu G., Wang D. Performance assessment and optimization of two novel cogeneration systems integrating proton exchange membrane fuel cell with organic flash cycle for low temperature geothermal heat recovery. Energy 2022;243:122725.

[39] Zhang B, Maloney D, Harun N F, Zhou N, Pezzini P, Medam A, Hovsapian R, Bayham S, Tucker D. Rapid load transition for integrated solid oxide fuel cell–Gas turbine (SOFC-GT) energy systems: A demonstration of the potential for grid response. Energy Conversion and Management 2022;258:115544.

[40] Zhuang Q, Geddis P, Runstedtler A, Clements B. An integrated natural gas power cycle using hydrogen and carbon fuel cells. Fuel 2017;209:76-84.

10

Hydrogen-powered Transportation Vehicles

Razzak Khan[1], Rishabh Kumar[1], Anuj Kumar[1], and Anoop Kumar Shukla[2]

[1]School of Mechanical Engineering (SMEC), Vellore Institute of Technology, India
[2]Amity School of Engineering and Technology, Amity University, India
E-mail: anuj.iitr02@gmail.com; razzak.khan2020@vitstudent.ac.in; rkr40209@gmail.com; shukla.anoophbti@gmail.com

10.1 Introduction

The most important problems of the 21st century are the world's need for more energy and the environmental problems that come with it, like global warming and climate change. It was acknowledged internationally that the rise of global temperature by 2 °C is inevitable (Álvarez Fernández, 2018). Due to this restriction, the design objectives for the current energy system must be expanded to include environmental effects, sustainability throughout its life span, constitutive materials, and hazardous products in addition to technology and economics (Hosseini, 2019). Transportation vehicles and machinery, which virtually solely rely on fossil fuels, utilize a considerable portion of the world's energy (Chapman, 2007). Energy analysts and environmentalists are emphasizing the need to adopt clean and sustainable energy resources due to the depletion of fossil fuel supplies and the rising pace of global greenhouse gas (GHG) emissions (Hosseini, 2016). Currently, fluid fossil fuels account for a significant share (65%) of the world's energy requirements due to their widespread availability and practical application. However, the production of fossil fuels will progressively decline over time (NejatVezirog, 2008). About half of the oil generated globally is consumed by the automobile sector. This energy consumption is expected to double or triple by 2050, and it seems improbable that the oil and gas supply will be able to keep up with this requirement. Across the globe, all countries are looking for environmentally friendly ways to meet their energy needs, particularly in the transportation industry (Hosseini, 2013). In many countries, hydrogen (H_2) and fuel cells (FCs) are viewed as crucial alternative sources of energy for developing clean and renewable energy systems for transportation, stationary electricity, industrial, and household use in the future (Edwards, 2008).

In the majority of governmental strategic plans, hydrogen (H_2) is regarded as one of the most significant sources to meet our future energy needs (Elnashaie, 2007). In internal combustion engines and fuel cells, hydrogen produces much fewer hazardous exhaust emissions, and on a mass basis, H_2 has a heating value three times greater than that of petrol (Fayaz, 2012). The advantages of using hydrogen in place of fossil fuels are innumerable. The fuel is regarded as environmentally beneficial because it solely emits water vapor when utilized in a fuel cell, which is one of its most significant advantages (Zeng, 2010). A viable solution for the transportation industry's future is the use of hydrogen as a fuel in vehicles that are powered by ICEs or fuel cells. The fuel production sector would be significantly altered by a hydrogen-based transportation system, which presents new economic opportunities and challenges (Chen, 2019). Due to the low density of H_2, certain obstacles arise, such as bulk storage, distribution, and on-board vehicle storage. However, the advantages of using it in transportation are sufficient to make hydrogen a viable alternative to the present transportation system. Commonly accessible hydrogen can be produced using renewable energy, but these sources are insufficient to supply the world's energy needs globally (Lindorfer, 2019). This is due to the absence of suitable, affordable hydrogen storage options currently. The use of hydrogen as an energy carrier in various sectors such as automobiles, industries, machinery, etc., to reduce the production of hazardous gases is still up for debate. However, currently, it is not economically viable (Butler, An overview of development and challenges in hydrogen-powered vehicles, 2019).

Due to numerous constraints in the production of hydrogen and its utilization process, examination of the development and many characteristics of hydrogen, particularly in the context of transportation engines, has emerged as one of the most important research areas. This book chapter reviews current advancements in hydrogen-based transportation engines with an emphasis on ICEs and fuel cell vehicles (FCVs) in order to assess the viability of a hydrogen-based transportation system.

10.2 Hydrogen-based Transportation Engines

There are two main ways to power vehicles: internal combustion engines that use hydrogen (H2ICE) and fuel cells that use hydrogen (HFCs) (Ehsani, 2018). The advantages of hydrogen-driven fuel cell vehicles (FCVs) are high efficiency, a lack of toxic emissions, silent operation, and a flexible design (Dicks, 2018). It generates energy using oxygen and hydrogen through electrochemical processes. Furthermore, the benefits of H2ICEs include

Table 10.1 Comparison of recent developed engines with fossil fuel-based engines (Butler, An overview of development and challenges in hydrogen powered vehicles, 2019).

Characteristic	Engine 1 Gasoline only	Engine 2	Engine 3	Engine 4	Engine 5	Engine 6	Engine 7
Engine make/model	GM 1.4 L Ecotec DOHC	F-150	113 kW AC Syn-chronous (PEMFC)	MeM3-245 (modi-fied)	Kirloskar, Model TV1 (modi-fied)	Jaling JH600 (modi-fied)	Hyundai ix35 (modified)
Fuel type	Gasoline multi-PFI	Diesel DI	CH_2 gas	Gasoline DL: HHO gas (5.9% H_2) DI	Diesel DI, CH_2 gas (max 20%) PFI	H_2 gas Spaced dual injection	CH_2 gas
Fuel storage: Method Tank volume (dm^3)	HPDE 34.1	HPDE 98	Type-IV CF	-	CH_2 at 130 bar-	CH_2 -	CH_2 at 700 bars
Classification Experimental (EX) or developed (DE)	DE	DE	DE	EX	EX	EX	EX
Orientation; No. of cylinders/No. of cells Displacement vol. Power output (KW) Torque (N-M)	In-line; 4-cylinder 1.399 73 127	V-6, 60ž- 2.99 186 596	-370 - 114 335	In-line; - 1.091 37.5 78.5	Single cylinder- 0.661 5.2 30	Single cylinder- - 30 51	-440 - 100 300
Availability # of vehicles sold in the US from 2015 to 2018	114,555	189,290	4,644	-	-	-	-
Efficiency: Fuel economy	33 mpg-	25 mpg-	66 mi/kg-	-33.3-	-27	-35-41 (at 1000-6000 rpm)	49 mi/kg-
Environmental GHG rating CO_2 emission (gram/mi; ppm)	8 273; -	4 412; -	10 0; 0	- -	- -; 35	- -; 0	10 0; 0

their dependence on a wide variety of sectors with significant production infrastructure, their potential to supply "flex-fuel" to ease the transition period, and their support for the implementation of the infrastructure for hydrogen (Verhelst 2014). HFC and electric vehicles (EVs) utilize hazardous materials; this results in limiting the use of such devices. Platinum is needed for fuel cells (FCs), and as demand rises, the price will increase. Battery-based electric vehicles (BEVs) utilize the earth's rare elements, which are challenging to generate in significant quantities. According to predictions, when compared to 1990s levels, FCVs can reduce GHG emissions in the US by 80% by the 2100s, and the nation's transportation system's reliance on gasoline fuel will be eliminated by the 2100s (Dougherty, 2009). In the last century, significant advancements in hydrogen-based ICES automobiles and hydrogen FCVs have been made. The comparison of recent hydrogen-based vehicles with several fuel-based engines is shown in Table 10.1.

10.3 Hydrogen Production Method

Several different feedstocks, including fossil and renewable resources, can be used to generate hydrogen. As shown in Figure 10.1, several process technologies, including chemical, biological, electrochemical, photochemical, and thermochemical, can be used to produce hydrogen (Riis, 2006). Some common techniques for producing hydrogen are discussed in subsequent sections.

10.3.1 Steam-methane reforming

Steam reforming is now among the most popular and economical ways to produce hydrogen (Ogden, 1999). There are mainly two steps for the entire process. In the first stage, the feedstock of hydrocarbons is supplied with steam into a tubular catalytic reactor. This process results in the production of syngas, a combination of carbon dioxide (CO_2) and hydrogen (H_2). When a fraction of the initial fuel (heated gas) inside the reactor burns, oxygen or air is injected as necessary to increase the reaction temperature. The second stage involves feeding the cooled product gas into the CO catalytic converter, which primarily uses steam to convert carbon monoxide into carbon dioxide and hydrogen (3), as shown in Figure 10.2. In this method, a sulfur-free raw material is necessary for the catalytic process in order to prevent the catalyst from becoming inactive. The SR technique is suitable for relatively low temperatures, such as more than 500 °C for the majority of conventional

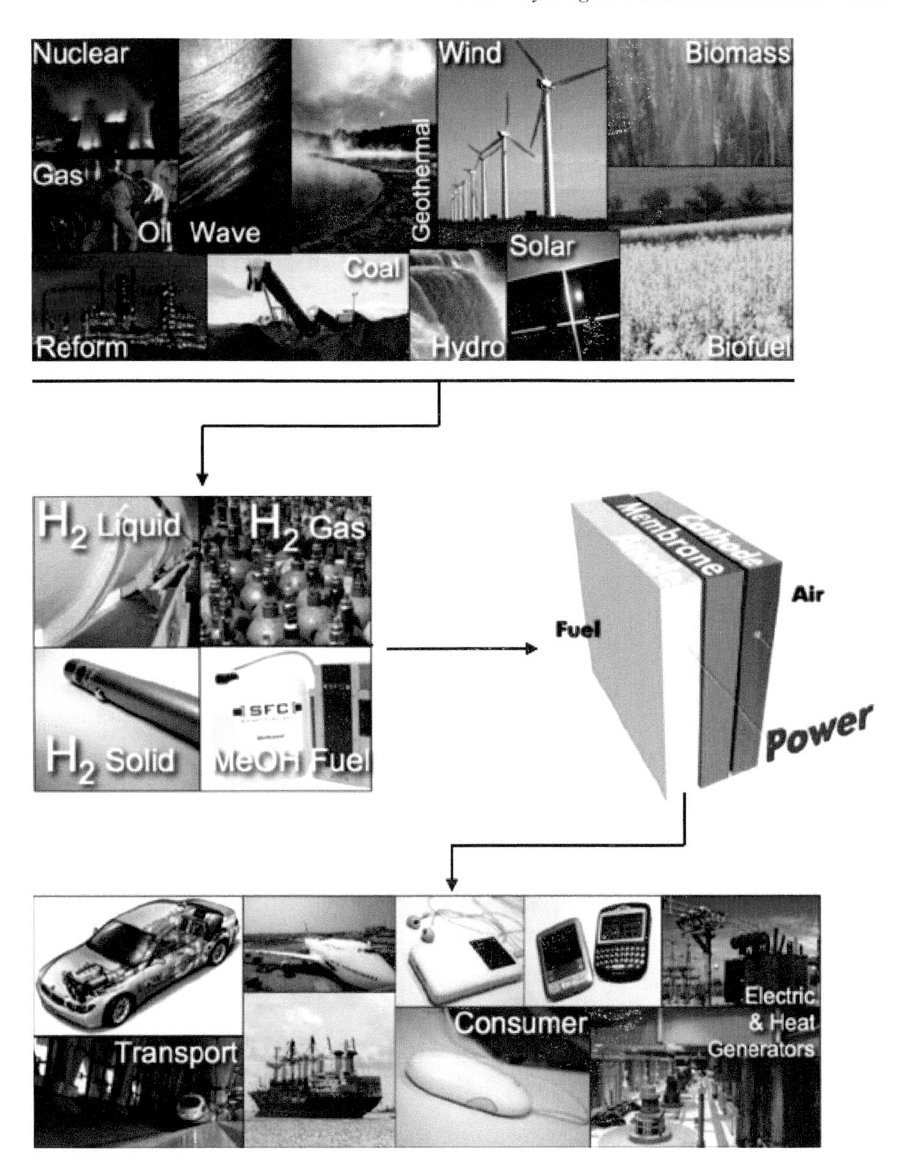

Figure 10.1 Multiple production processes and energy sources are connected to a wide range of fuel cell applications using hydrogen (Edwards PP, 2008).

hydrocarbons and 180 °C for methanol and oxygenated hydrocarbons (Farrauto, 2003). On an industrial scale, the thermal efficiency of hydrogen generation by the steam reforming method ranges between 70% and 85% (Sorensen, 2011).

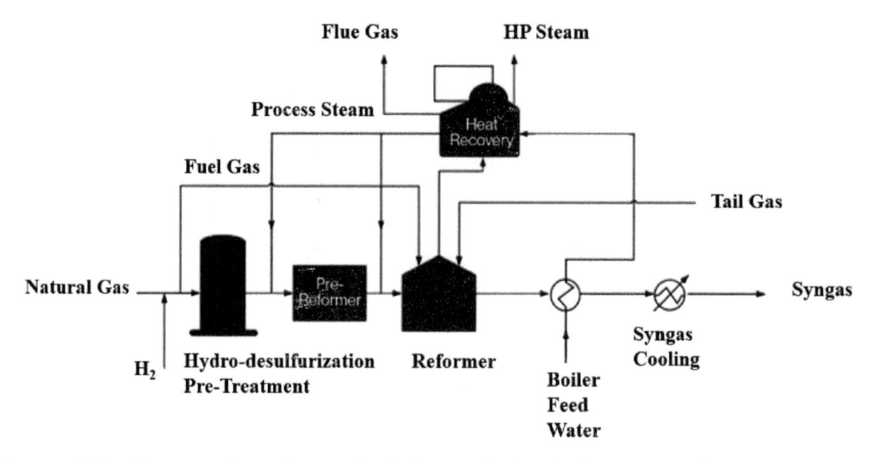

Figure 10.2 Steam reformation method for producing hydrogen gas (Image courtesy: https://www.mvsengg.com/blog/steam-methane-reformer/ accessed on 27/03/2023).

10.3.2 Gasification of coal

"Gasification" is also known as the "partial oxidation process," which produces H_2 from a number of hydrocarbon fuels such as coal, heavy-waste oil, and refined petroleum products. The primary methods for producing hydrogen are coal gasification, heavy crude oil, petroleum coke, and reformation of heavy oil and natural gas (Zeng, 2010). Basically, gasification

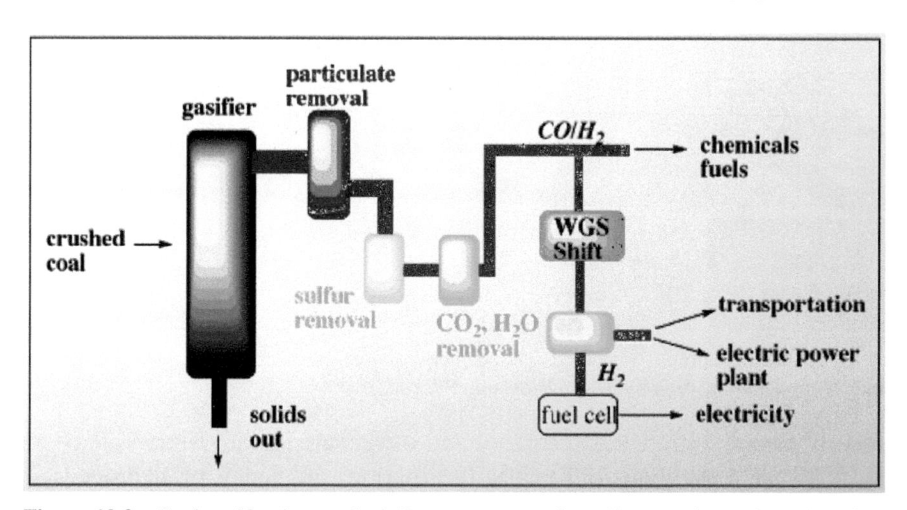

Figure 10.3 Coal gasification method (Image courtesy: http://butane.chem.uiuc.edu/pshapley/environmental/l5/1.html/accessedon27/01/2023).

is the reaction of solid fuels with air, oxygen, steam, carbon dioxide, or a combination of these gases at temperatures above 800 °C in a reducing environment where the air-to-oxygen ratio is controlled. The chemical bonds in the molecular structure of coal are broken by heat and pressure, which sets off chemical reactions with the steam and oxygen. As illustrated in Figure 10.3 , result comes out to be a gas, which can be further used as a source of energy or as a raw material (syngas) to make chemicals, liquid fuels, or other gaseous fuels. This is called "synthesis gas" or "syngas" and is mostly made up of carbon monoxide and hydrogen. The problem associated with this process is its low thermal efficiency because the feedstock must be heated to dry out the water. Large-scale gasification reactors need to be continuously fed with enormous volumes of feedstock. Due to their lower heating value, they can reach thermal efficiencies of around 35%–50% (Holladay, 2009).

10.3.3 Electrolysis

Water electrolysis could be a promising technique for hydrogen production in the future. Currently, just 4% of the hydrogen produced worldwide comes from this method (Konieczny, 2008). The process of electrolysis involves transmitting a current directly between two electrodes that are submerged in water, which causes the breakdown of chemical bonds between liquid molecules, thus releasing H_2 and O_2 as shown in Figure 10.4. The process

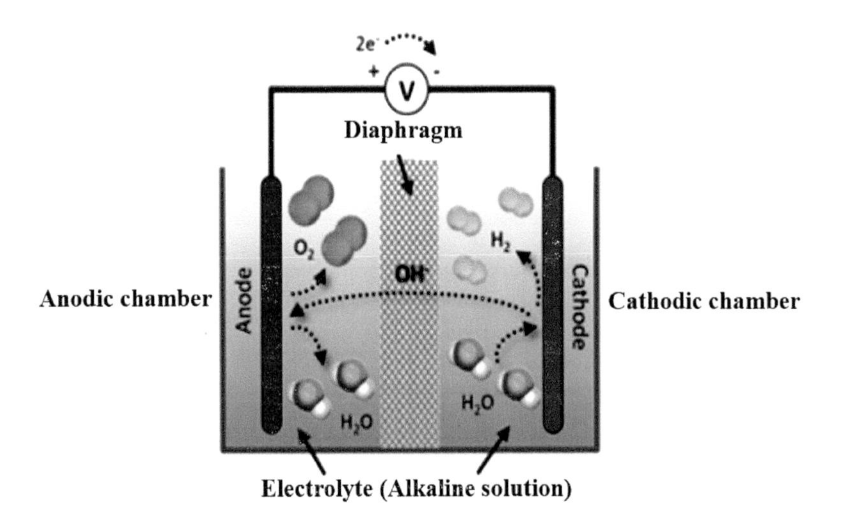

Figure 10.4 Process of electrolysis (Amores, Sánchez, Rojas, & Sánchez-Molina, 2021).

of electrolysis is conducted at room temperature, and sulfuric acid is used as an electrolyte. This process is environmentally friendly because no GHG emissions are produced during the process, and the produced oxygen has further industrial uses. But compared to the methods previously discussed, electrolysis is the most promising technique to produce hydrogen. Currently, electrolysis is carried out using electrolyzers with a power range of a few kilowatts to 2000 kW. Even though proton exchange membranes and solid-oxide electrolytic cells have been discovered, alkaline-based electrolysis techniques are most popular (Grigoriev, 2006). The SOEC technology faces difficulties with chromium diffusion, heat transfer, corrosion, and sealing. PEM electrolyzers are more effective than alkaline and free from corrosion problems, like in the case of SOEC. However, the cost of an alkaline-based electrolyzer system is lower than that of PEM.

10.3.4 Photo-electrolysis

This process is one of the renewable ways to make hydrogen, and it works well and costs little. However, it is in its development phase (Faaij, 2002). At present, it is the most affordable and efficient technique for producing hydrogen from renewable energy sources. The photoelectrode consists of a semi-conductive device that captures solar energy while generating the required voltage to produce H_2 and O_2 with the help of breaking chemical bonds between water molecules directly. A photoelectrochemical (PEC)

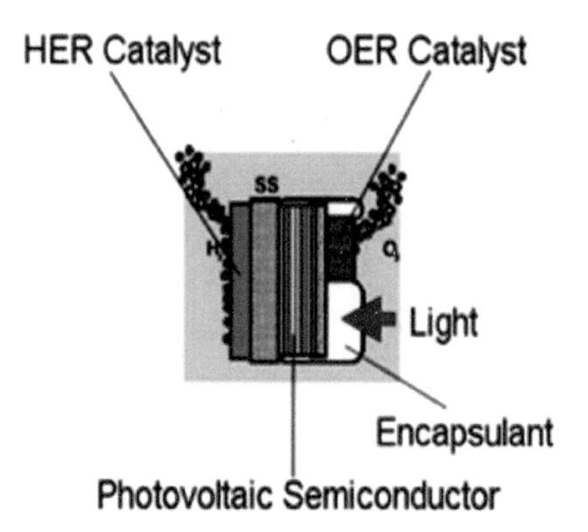

Figure 10.5 Generation of hydrogen from photo electrolysis (Image courtesy: https://bhuang02.tripod.com/photoelectrolysis.htm/accessedon27/01/2023).

energy storage system is used in photoelectrolysis to energize the electrolysis of water. Photoelectrodes also provide enough electrical power to enable the reactions of hydrogen and oxygen if they are immersed in an electrolyte solution and exposed to sunlight. For hydrogen production, electrons are discharged into the electrolyte, whereas free electrons are needed for oxygen generation. The efficiency of the electrolysis is also influenced by the PEM cell's catalytic layers, which necessitate the use of appropriate catalysts for splitting water. In this process, two catalysts were used, namely the hydrogen evolution reaction catalyst and the oxygen evolution reaction catalyst, as shown in Figure 10.5. An enclosing layer is an essential component of a photoelectrode that can prevent the oxidation of semiconducting materials within the electrolyte solution. This layer should be sufficiently transparent in order to absorb maximum solar energy and allow it to penetrate the photovoltaic layer.

10.3.5 Hydrogen from biomass

Besha (2020) developed a technical pathway known as biomass gasification, which converts biomass into H_2 and other products without burning using a controlled process involving heat energy, steam, and oxygen. The carbon emissions associated with this technology may be low due to the removal of CO_2 from the atmosphere during biomass growth, especially if carbon capture, utilization, and storage are used in the long run. Building and operating gasification facilities for biofuels can provide insights and lessons about producing hydrogen. A promising approach for generating renewable hydrogen has been suggested: biomass gasification. Utilizing biomass

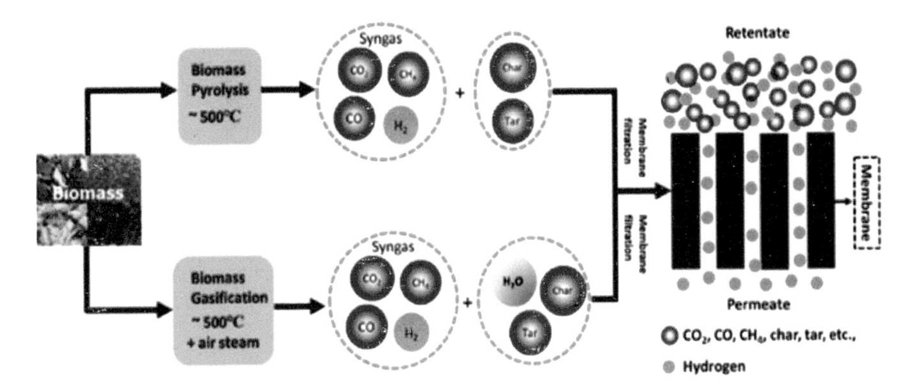

Figure 10.6 Production of hydrogen gas from biomass gasification (Besha, 2020).

resources will help to create a very effective and clean method for producing vast amounts of hydrogen. Consequently, there is less reliance on unsafe fossil fuels (Balat, 2009). Thermochemical and biochemical processes are the two categories of hydrogen energy conversion techniques. Typically, thermochemical processes are less expensive because they can attain a higher reaction rate when operated at high temperatures. Figure 10.6 shows how to produce pure hydrogen fuel, this process uses gasification or pyrolysis techniques (the burning of biomass without oxygen).

10.3.6 Hydrogen from nuclear power

Nuclear power reactors can produce hydrogen in a number of ways that would significantly reduce air pollutants while utilizing the consistent thermal energy and dependable electricity it consistently produces. Figure 10.7 displays a schematic diagram related to production of hydrogen from nuclear reactor which include different chemical processes like water electrolysis utilizing nuclear energy and the iodine-sodium cycle. Additionally, high-temperature electrolysis also makes use of the heat from nuclear power plants to reduce the amount of electricity needed for electrolysis. Existing nuclear power stations have the ability to generate high-quality steam at a cheaper

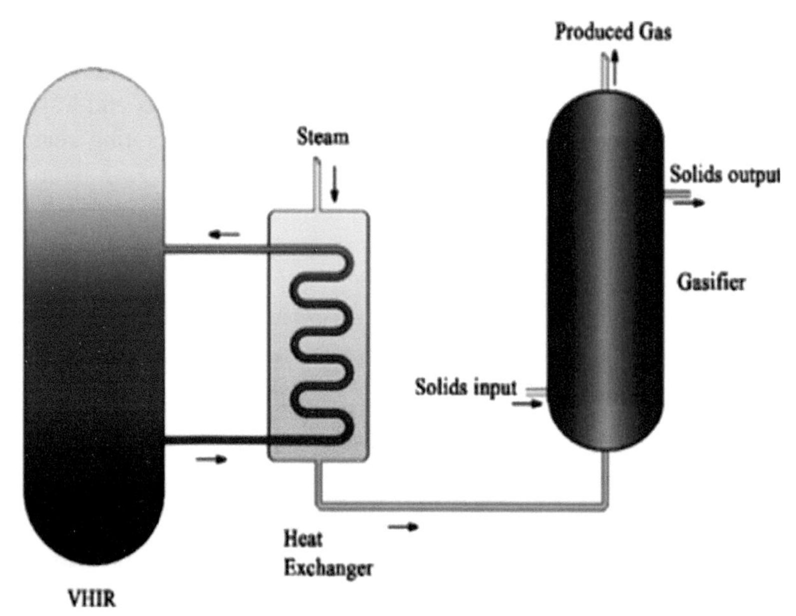

Figure 10.7 Hydrogen generation from nuclear reactor (Belghit*, 2020).

cost than natural gas boilers, which can be used in a wide range of industrial processes, namely steam reforming. However, when this improved steam is electrolyzed and separated into pure H_2 and O_2, the case for nuclear power becomes even more convincing. More than 150,000 tons of hydrogen may be produced annually by a single 1000-MW nuclear reactor.

10.4 Challenges Involving Hydrogen-integrated Vehicles

10.4.1 Safety of hydrogen-based vehicles

Basically, fuels with a high heat capacity, a high coefficient of diffusion, and a low density are generally regarded as safe. Some characteristics, such as broader combustion limits and a low ignition temperature, reduce the danger of a particular fuel by broadening the potential fire circumstances. The fuel is less safe when the temperature of the flame and explosive energy is higher because the fire turns out to be more destructive. Since hydrogen is so light (around 6.9% of the air density), there is almost no chance of combustion; it is four times more diffuse than nitrous oxide and 12 times more diffuse than gasoline (Sharma, 2015). Since hydrogen has no negative attributes, its release will not have an adverse effect on the environment. Since hydrogen cannot easily be made to explode, a spark in a hydrogen/air mixture would cause a quick fire to start (Hawkins, 2006). Safety concerns arise due to the possibility that lower hydrogen concentrations might ignite (Midilli, 2005).

For system storage and safety, the US DOE has established some standards. The system must comply with SAE J2579 for system safety in terms of penetration and leakage; for toxicity, the system should comply with applicable regulations and undergo failure analysis before being considered safe. Instead of testing individual components or stored materials separately, the penetration and leakage tests are performed on the entire storage system. The safety guidelines cover the transportation system, vehicle production, certification, operation, fuel distribution, and end-of-life challenges, all of which must adhere to relevant federal, state, and local regulations.

10.4.2 Storage of hydrogen

Right now, storing hydrogen is the biggest problem, and it needs to be fixed before a hydrogen-based fuel system can be set up that is both technically and economically viable (Balla, 2009). In general, there are two methods to run H_2-powered vehicles on the road. First, in the case of an IC engine, hydrogen is rapidly burned with oxygen present in the air. Second, in a fuel cell, hydrogen is electrochemically “burned” with oxygen from the air, producing heat and powering an electric engine (Schlapbach Louis, 2001). Due to the low density of hydrogen, its enormous on-board storage tank is the biggest drawback to its use as a transportation fuel (M. B., 2008). Finding a storage medium that satisfies three essential criteria is the immediate challenge, which include: (1) a high density of hydrogen; (2) to be compatible with the current generation of FCs, the release and charge cycle must be reversible at mild temperatures in the range of 70–100 °C; and (3) hydrogen release and charge kinetics that are fast and have low energy barriers. The first requirement is that the tank’s material be chemically strong and tightly packed atomically. And the second material requires weak hydrogen bonds, which are likely to break at moderate temperatures. The third requires atomic packing that is free to enable rapid diffusion of hydrogen between the bulk and the surface. To minimize the deterioration induced by heat rejection during hydration, the material must have adequate thermal conductivity (Sharma, 2015). By altering the state conditions, hydrogen can be physically stored due to its light weight, simplicity, and easy availability, and can be chemically stored in solid and liquid forms (Shakya, 2005).

The different types of hydrogen storage are illustrated in a flowchart (Figure 10.8). Transportation and storage of hydrogen pose significant research challenges for materials. Because they do not meet material requirements, many popular bulk materials have already been considered as storage options but rejected. However, nanoscience presents new possibilities for overcoming this difficulty.

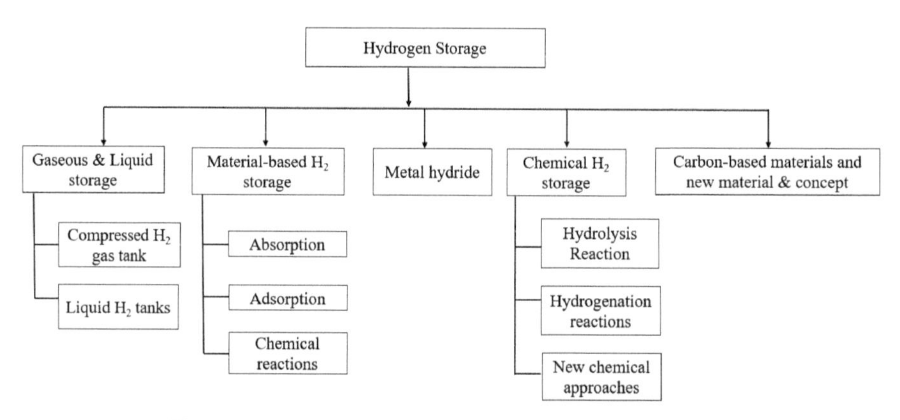

Figure 10.8 Flowchart showing hydrogen storage methods.

10.5 H_2-integrated Internal Combustion Engines

Internal combustion engines are a key component of the transportation system. However, they consume a lot of liquid fossil fuels like gasoline and diesel. When petroleum fuel is used to produce energy in ICEs, several pollutants are released into the atmosphere, including carbon monoxide (CO), nitrogen oxides (NOx), unburned hydrocarbons (UHCs), particulate matter (PM), and greenhouse gases. In comparison to gasoline engines, hydrogen-based ICES have a higher energy content, and their chemical and physical properties enable them to run more efficiently on extremely lean mixtures. There are mainly two types of hydrogen-based IC engines, namely: (1) pressure booster IC engines and (2) liquid hydrogen IC engines (Figure 10.9).

In a pressure-booster IC engine, the mixture of high-pressure hydrogen and air gas passing through the pressure regulator is compressed in the combustion chamber of the engine cylinder to enhance the power density of IC engines. In that case, hydrogen pressure is raised to reach volumetric hydrogen efficiencies that are comparable to those of liquid fuels (petrol and diesel). In the liquid hydrogen system, liquefied hydrogen is introduced into an expansion chamber, where it is transformed into a cold hydrogen gas before being carried to the combustion chamber. Hydrogen has a larger diffusivity, a significantly lower combustion power (0.02 MJ), and a higher ignition temperature in comparison with other fuels. These properties of hydrogen are desirable for a spark-ignition engine. Because hydrogen burns at a flame speed that is 5–10 times faster than that of diesel and methane, hydrogen-based spark-ignition engines can operate with less cyclic variation. In these types of engines, hydrogen is employed in one of the ways listed below.

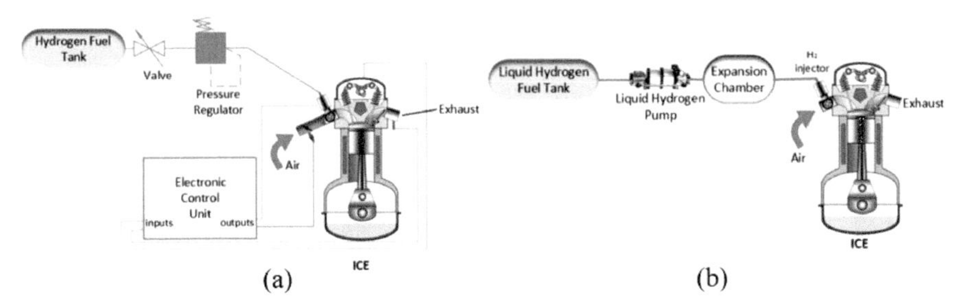

Figure 10.9 (a) Pressure booster IC engine. (b) Liquid H_2 IC engine (Gurz, 2017).

- **Manifold induction**: Injection of low-temperature H_2 into the manifold occurs through a tube that is under the control of a valve.
- **Direct introduction**: Hydrogen is stored in cryogenic cylinders. A pump evaporates liquid H_2 in a heat exchanger, and then it pumps cold H_2 gas into the engine. Pre-ignition is prevented, and nitrogen oxide generation during combustion is decreased, by employing cold hydrogen gas.
- **Adding hydrogen to gasoline**: In this method, a mixture of compressed hydrogen and gasoline is injected into the IC engine and set on fire with the help of a spark.

Additionally, hydrogen may be used in a compressed-ignition engine. It has been discovered that a CI engine can generate twice as much power as a similar engine operating in a premixed system (Ganesan, 2012). In hydrogen-fueled CI engines, injectors are used to deliver pressurized hydrogen gas into the engine cylinder. Since the injector nozzle controls the system's intake of pressurized hydrogen, the design of the injector is equally as critical as the design of the engine assembly. In H_2-based CI engines, they demonstrated a considerable decrease in greenhouse gas emissions, and under ideal conditions, it can reach over 50%. Heat release rate (HRR) and brake thermal efficiency both appear to have a significant increase under high load conditions when employing higher amounts of H_2 in an IC engine.

10.6 Fuel Cell Integrated Transportation Engines

Currently, the complexity of the energy sector's structure makes it challenging to integrate hydrogen and fuel cells (HFCs) into the world's energy system. HFCs are stationary devices that turn hydrogen and oxygen into electricity through an electrochemical process. If H_2 and O_2 are supplied into the system, HFCs will produce electricity. Fuel cells (FCs) are less efficient than batteries, with ratings of roughly 50% at lower heating values. However, compared to a battery's similar electrical capacity, an HFC's volumetric energy density is substantially higher. FCs are different from batteries because the FC is not a storage device. It can store hydrogen for an extended period of time because it does not self-discharge the hydrogen stored in it. The dissociation of power and energy ratings in hydrogen fuel cells is a distinctive and alluring feature. Since these are proportionate in a battery, battery-electric vehicles (BEVs) with a long range are costly, heavy, and extremely powerful. They outperform BEVs in terms of power availability because the FC identifies the power rating, whereas the storage of hydrogen determines the length of the energy production, equivalent to an IC engine. There have been numerous experimental studies with several types of FCs in hydrogen-powered engines that have been evaluated, and the underlying idea behind how they work is identical. The main parts of the fuel cell-based engines are the anode, cathode, and electrolyte. In that, the supply of hydrogen takes place at the anode, and at the cathode, oxygen is charged. At the anode, a catalyst speeds up the process of electron separation from hydrogen atoms. The cathode and an associated device are powered by the electrons traveling along a route, and while the produced protons are passing through a proton-conducting electrolyte, oxygen is being reduced. Protons and oxygen mix to generate water on the cathode side. The HFC system typically uses oxygen and hydrogen to produce energy and water without harmful pollutants. The list of fuel cell-based vehicles available on the global market is shown in Table 10.2.

There are mainly two types of fuel cells, i.e., polymer electrolyte membrane (PEM) fuel cells and alkaline membrane fuel cells (AMFCs), as shown in Figure 10.10. Both PEM and AMF fuel cells work in the same way, but PEMFCs have an acid membrane and AMFCs have an alkaline membrane. AMFCs are less expensive than PEMFCs because, at the anode, they could use a variety of different metal catalysts and cathodes, which are more cost-effective. The membrane's life, power density, corrosion, ability to handle higher temperatures, water management, and electrocatalyst at the anode are some of the problems with this type of fuel cell (Alesker, 2016).

PEMFC is the second kind of FC; it operates at temperatures below 120 °C and uses perfluoro-sulfonic acid as the charge carrier substance. They can start up more quickly as compared with different types of FCs, making them perfect for use in transportation applications. To make the cell last longer and control how water moves through it, the amount of precious metals used to make PEMFCs, especially platinum, must be cut down or eliminated. Flooding is a common problem with PEMFC operation. It happens when

Table 10.2 Hydrogen-based vehicles available globally.

No.	Model	Commencement of production
1	Hyundai ix35 fuel cell	2013 to present
2	Hyundai Tuscon FCEV	2014 to present
3	Toyota Mirai	2014 to present
4	Honda Clarity	2016 to present
5	Hyundai Nexo	2018 to present

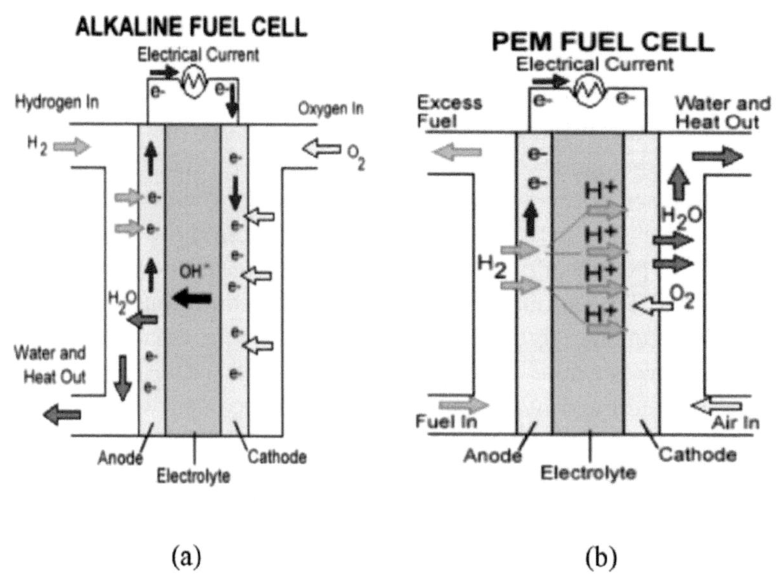

(a) (b)

Figure 10.10 Different types of fuel cells (Hames, 2018).

the humidity of the gas rises, which speeds up the rate at which platinum dissolves and the carbon support corrodes.

10.7 Future Possibilities for using Hydrogen-based Transportation Systems

The potential long-term limitations on greenhouse gas emissions, the price of hydrogen, and the rate of development of various hydrogen-based technologies all affect the future hydrogen market. Hydrogen has numerous social, environmental, and economic advantages that make it a viable clean energy source for the next generation of transportation systems. Three significant obstacles must be overcome in order to progress technologically and economically toward a hydrogen-powered transportation system. First, with the introduction of hydrogen on par with other fuels like gasoline, the cost of its manufacturing and delivery needs to be significantly reduced. The cost of hydrogen mainly depends on some factors, such as the method used to produce it, the major energy source, and the model used. Second, to ensure adequate driving range, an improved technique to store hydrogen for transportation vehicles needs to be discovered. The low cost of hydrogen in comparison to other fuels, as well as technological advancements in the design of vehicle engines toward hydrogen-powered engines, will have a significant impact on the use of hydrogen as a fuel in the transportation system in the future. Additionally, social–cultural aspects are vital for the advancement of hydrogen in the transportation sector since they affect how present conventional fossil fuel-based engines are controlled in terms of mobility and household appliances. Informational campaigns have been discovered to have a substantial impact on the advancement of hydrogen technologies. The main focus of a campaign must be on the opinions and attitude of an individual having interest and little bit of knowledge regarding hydrogen powered vehicles. (Elnashaie, 2007). Future market growth for hydrogen is mostly dependent on four factors: (1) the future hydrogen price; (2) the rate at which various hydrogen-based technologies are developing; (3) possible long-term limitations on greenhouse gas emissions; (4) the price of alternative energy systems. Hydrogen could reduce carbon and other harmful emissions from the transportation industry and decrease reliance on foreign oil in the future.

References

[1] Alazemi, J., & Andrews, J. (2015). Automotive hydrogen fuelling stations: An international review. *Renewable and sustainable energy reviews*, *48*, 483-499.

[2] Alesker, M., Page, M., Shviro, M., Paska, Y., Gershinsky, G., Dekel, D. R., & Zitoun, D. (2016). Palladium/nickel bifunctional electrocatalyst for hydrogen oxidation reaction in alkaline membrane fuel cell. *Journal of Power Sources*, *304*, 332-339.

[3] Balat, M. (2008). Potential importance of hydrogen as a future solution to environmental and transportation problems. *International journal of hydrogen energy*, *33*(15), 4013-4029.

[4] Balat, M., & Balat, M. (2009). Political, economic and environmental impacts of biomass-based hydrogen. *International journal of hydrogen energy*, *34*(9), 3589-3603.

[5] Ball, M., & Wietschel, M. (2009). The future of hydrogen–opportunities and challenges. *International journal of hydrogen energy*, *34*(2), 615-627.

[6] Boretti, A. (2011). Advantages of the direct injection of both diesel and hydrogen in dual fuel H2ICE. *international journal of hydrogen energy*, *36*(15), 9312-9317.

[7] Bradley, D., Lawes, M., Liu, K., Verhelst, S., & Woolley, R. (2007). Laminar burning velocities of lean hydrogen–air mixtures at pressures up to 1.0 MPa. *Combustion and Flame, 149*(1-2), 162-172.

[8] Hosseini, S. E., & Butler, B. (2020). An overview of development and challenges in hydrogen powered vehicles. *International journal of green energy, 17*(1), 13-37.

[9] Chapman, L. (2007). Transport and climate change: a review. *Journal of transport geography, 15*(5), 354-367.

[10] Chen, S., Kumar, A., Wong, W. C., Chiu, M. S., & Wang, X. (2019). Hydrogen value chain and fuel cells within hybrid renewable energy systems: Advanced operation and control strategies. *Applied Energy, 233*, 321-337.

[11] Dicks, A. L., & Rand, D. A. (2018). *Fuel cell systems explained.* John Wiley & Sons.

[12] Dougherty, W., Kartha, S., Rajan, C., Lazarus, M., Bailie, A., Runkle, B., & Fencl, A. (2009). Greenhouse gas reduction benefits and costs of a large-scale transition to hydrogen in the USA. *Energy policy, 37*(1), 56-67.

[13] Edwards, P. P., Kuznetsov, V. L., David, W. I., & Brandon, N. P. (2008). Hydrogen and fuel cells: towards a sustainable energy future. *Energy policy, 36*(12), 4356-4362.

[14] Ehsani, M., Gao, Y., Longo, S., & Ebrahimi, K. M. (2018). *Modern electric, hybrid electric, and fuel cell vehicles.* CRC press.

[15] Elnashaie, S., Chen, Z., & Prasad, P. (2007). Efficient production and economics of clean-fuel hydrogen. *International journal of green energy, 4*(3), 249-282.

[16] Farrauto, R., Hwang, S., Shore, L., Ruettinger, W., Lampert, J., Giroux, T., ... & Ilinich, O. (2003). New material needs for hydrocarbon fuel processing: generating hydrogen for the PEM fuel cell. *Annual Review of Materials Research, 33*(1), 1-27.

[17] Fayaz, H., Saidur, R., Razali, N., Anuar, F. S., Saleman, A. R., & Islam, M. R. (2012). An overview of hydrogen as a vehicle fuel. *Renewable and Sustainable Energy Reviews, 16*(8), 5511-5528.

[18] Fernández, R. Á., Caraballo, S. C., Cilleruelo, F. B., & Lozano, J. A. (2018). Fuel optimization strategy for hydrogen fuel cell range extender vehicles applying genetic algorithms. *Renewable and sustainable energy reviews, 81*, 655-668.

[19] Frenette, G., & Forthoffer, D. (2009). Economic & commercial viability of hydrogen fuel cell vehicles from an automotive manufacturer perspective. *International Journal of Hydrogen Energy*, *34*(9), 3578-3588.

[20] Ganesan, V. (2012). *Internal combustion engines*. McGraw Hill Education (India) Pvt Ltd.

[21] Grigoriev, S. A., Porembsky, V. I., & Fateev, V. N. (2006). Pure hydrogen production by PEM electrolysis for hydrogen energy. *International Journal of Hydrogen Energy*, *31*(2), 171-175.

[22] Hamelinck, C. N., & Faaij, A. P. (2002). Future prospects for production of methanol and hydrogen from biomass. *Journal of Power sources*, *111*(1), 1-22.

[23] Hawkins, S., & Joffe, D. (2006). Technological characterisation of hydrogen storage and distribution technologies. *Policy Studies Institute: London, UK*.

[24] Holladay, J. D., Hu, J., King, D. L., & Wang, Y. (2009). An overview of hydrogen production technologies. *Catalysis today*, *139*(4), 244-260.

[25] Hosseini, S. E., Andwari, A. M., Wahid, M. A., & Bagheri, G. (2013). A review on green energy potentials in Iran. *Renewable and Sustainable Energy Reviews*, *27*, 533-545.

[26] Hosseini, S. E., & Wahid, M. A. (2016). Hydrogen production from renewable and sustainable energy resources: Promising green energy carrier for clean development. *Renewable and Sustainable Energy Reviews*, *57*, 850-866.

[27] Hosseini, S. E. (2019). Development of solar energy towards solar city Utopia. *Energy Sources, Part A: Recovery, Utilization, and Environmental Effects*, *41*(23), 2868-2881.

[28] Hosseini, S. E., & Butler, B. (2020). An overview of development and challenges in hydrogen powered vehicles. *International journal of green energy*, *17*(1), 13-37.

[29] Konieczny, A., Mondal, K., Wiltowski, T., & Dydo, P. (2008). Catalyst development for thermocatalytic decomposition of methane to hydrogen. *International Journal of Hydrogen Energy*, *33*(1), 264-272.

[30] Lindorfer, J., Reiter, G., Tichler, R., & Steinmüller, H. (2019). Hydrogen fuel, fuel cells, and methane. In *Managing Global Warming* (pp. 419-453). Academic Press.

[31] Lipman, T. (2011). An overview of hydrogen production and storage systems with renewable hydrogen case studies. *Clean Energy States Alliance*, *32*.

[32] Midilli, A., Ay, M., Dincer, I., & Rosen, M. A. (2005). On hydrogen and hydrogen energy strategies: I: current status and needs. *Renewable and sustainable energy reviews*, *9*(3), 255-271.

[33] Ogden, J. M., Steinbugler, M. M., & Kreutz, T. G. (1999). A comparison of hydrogen, methanol and gasoline as fuels for fuel cell vehicles: implications for vehicle design and infrastructure development. *Journal of power sources*, *79*(2), 143-168.

[34] Riis, T., Hagen, E. F., Vie, P. J., & Ulleberg, Ø. (2006). Hydrogen production and storage R&D priorities and gaps. *International Energy Agency-Hydrogen Co-Ordination Group-Hydrogen Implementing Agreement.*

[35] Schlapbach, L., & Züttel, A. (2001). Hydrogen-storage materials for mobile applications. *nature*, *414*(6861), 353-358.

[36] Singh, S., Jain, S., Venkateswaran, P. S., Tiwari, A. K., Nouni, M. R., Pandey, J. K., & Goel, S. (2015). Hydrogen: A sustainable fuel for future of the transport sector. *Renewable and sustainable energy reviews*, *51*, 623-633.

[37] Sorensen, B. (2011). Hydrogen and fuel cells: emerging technologies and applications.

[38] Verhelst, S. (2014). Recent progress in the use of hydrogen as a fuel for internal combustion engines. *international journal of hydrogen energy*, *39*(2), 1071-1085.

[39] Zeng, K., & Zhang, D. (2010). Recent progress in alkaline water electrolysis for hydrogen production and applications. *Progress in energy and combustion science*, *36*(3), 307-326.

11

Hydrogen Utilization for Renewable Ammonia Production (Power-to-Ammonia)

Alper Can Ince[1,2*], C. Ozgur Colpan[3], Mustafa Fazıl Serincan[4], and Ugur Pasaogullari[1,2]

[1]Center for Clean Energy Engineering, University of Connecticut, United States
[2]Department of Mechanical Engineering, University of Connecticut, United States
[3]Dokuz Eylul University, Faculty of Engineering, Mechanical Engineering Department, Izmir
[4]Gebze Technical University, Faculty of Engineering, Mechanical Engineering Department, Turkey
E-mail: alper_can.ince@uconn.edu, ozgur.colpan@deu.edu.tr, mfserincan@gmail.com, ugur.pasaogullari@uconn.edu

Abstract

Ammonia is a liquid and inorganic chemical that is simply storable and transportable as industrial raw material. Ammonia has been extensively used for various applications over a century from producing agricultural fertilizers (e.g., ammonium nitrate, ammonium phosphate, and urea) to producing power through an internal combustion engine (ICE) with a high-octane rate of 110–130 and the fuel cells (e.g., ammonia fuel cells). Moreover, ammonia has gained more attention in the last decade to be considered as a hydrogen carrier due to significant challenges in hydrogen storage and transportation. Even though the ammonia synthesis industry is still dominated by fossil fuels, which hurt the environment dramatically, renewable ammonia synthesis pathways can be a promising method to meet the ammonia requirement of the world. Therefore, the scope of this book chapter is to introduce the hydrogen-based renewable ammonia synthesis pathways by presenting worldwide simulative and demonstrative projects. Moreover, a thermodynamic analysis performed for the 1-MW power-to-ammonia system and the remarkable results taken from the analysis are discussed. Finally, the conclusions and future directions are presented.

Keywords: Hydrogen, green ammonia, water electrolyzer, power-to-ammonia, thermodynamic analysis

11.1 Introduction

The use of fossil fuels as the primary source of energy has led to climate change due to global warming, with temperatures possibly rising above 2 °C in the next three decades and beyond 4–6 °C within the next several decades [1–4]. Moreover, a considerable depletion of fossil fuels (due to industrialization and an increase in world population growth) and high dependency on geography have led to energy security problems (e.g., oil supply disruption and emerging gas security challenges) [5]. Renewable energy is considered the main solution to maintaining global temperatures below 2 °C [1]. Therefore, large-scale deployment of renewable power technologies (e.g., wind, solar PV, concentrating solar power, biomass, etc.) is necessary. However, the fluctuation in renewable resources that depend on, for example, solar radiation or wind velocity makes the need for energy storage technologies necessary. Here, energy storage technologies can meet energy demand when solar irradiation and wind are insufficient for storing surplus energy. To date, various mechanical, electrical, thermal, and chemical energy storage techniques have been developed to store extra electricity [6, 7]. The basic comparison among the storage techniques can be found in the literature. The chemical energy storage technique, which stores electricity in a chemical form (mostly fuel), is a sufficiently flexible mechanism that provides large quantities of energy to be stored over long periods at independent locations [8]. Energy can be stored in a chemical form, such as hydrogen or carbon-neutral hydrogen derivatives [9, 10]. Hydrogen and ammonia, which are feedstock chemicals, can be considered carbon-free fuel sources (they contribute no CO_2 emissions).

Hydrogen:

Hydrogen is an effective energy vector with an energy-rich substance that facilitates energy storage without releasing greenhouse gases that contribute to global warming [11]. From an economic point of view, hydrogen is currently produced from renewable sources at a cost of about $5 per 1 kg. As a frame of reference, 1 gallon of gasoline contains approximately the same energy as 1ă kg of hydrogen. The U.S. Department of Energy has launched a Hydrogen Shot that aims to achieve \$1 per 1-kg hydrogen in a decade (www.energy.gov). Therefore, it is projected to decrease the cost of hydrogen production from renewable energy to \$1/kg by 2031. However, hydrogen is a highly compressible gas, which leads to the need for extreme compression at a pressure of 70 MPa (in gaseous form) or a temperature of −250 °C (in liquid form) for the use of hydrogen for energy storage [12]. Maintaining this pressure or temperature is prohibitively expensive, which makes alternative hydrogen utilization approaches necessary. The use of ammonia as a potential hydrogen carrier for hydrogen delivery or off-board hydrogen storage (energy storage) has gained popularity in the last few years. For instance, over 182 days, the cost of ammonia storage would be 0.54 \$/kg-H_2, compared to a pure hydrogen storage cost of 14.95 \$/kg-H_2.

Ammonia:

Ammonia (NH_3) is the second most produced industrial chemical (it reached 146 million tons in 2016 [13]), and the production process has been intensively evolving over a period of a century. Ammonia is a strong candidate for flexible energy storage on the largest scale. Ammonia is easier, has more handling and distribution capacity, and has better commercial viability with a wide range of end uses (as shown in Figure 11.1). The main benefit

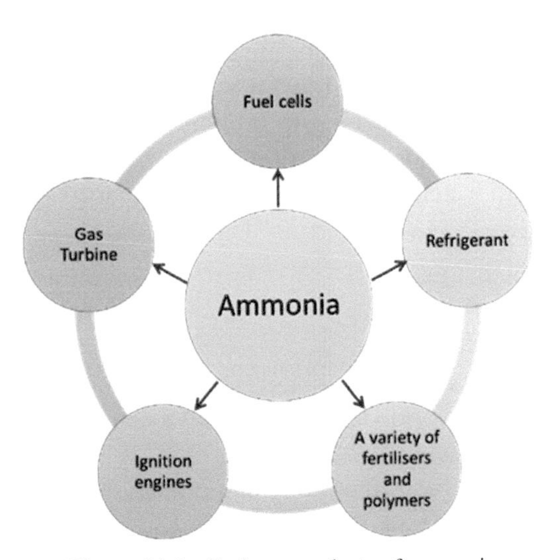

Figure 11.1 End-use products of ammonia.

of ammonia is that it can be used as a fuel for power production through fuel cells, gas turbines, and ignition engines.

From the point of view of fuel, ammonia has the following advantages:

- carbon-free substance;
- high octane numbers (110–130);
- the compactness of fuel tanks on vehicles;
- cost-effectiveness.

Through the combustion of ammonia fuel in ignition engines, the driving cost for a 100-km distance is $3.2, and the volume of the ammonia tank is found to be 18ă L. These results are comparable to the price compared to common gasoline vehicles that consume 6–7 L per 100 km, leading to a cost of $6–7 per 100 km. However, the use of ammonia still has the following disadvantages:

- Fossil fuels heavily dominate the current ammonia production methods, accounting for 1%–2% of the annual global energy demand and releasing about 2.9 metric tons of CO_2 per metric ton of NH_3.
- Power output is from small to utility-scale size, typically in the range of 0.1–1.0 MW.
- There is a risk of explosion due to the sudden release of ammonia [14].
- The reduction of NOx emissions and unburned ammonia, contaminants that directly impact climate change and are toxic to life, respectively. It is found that NO_2 may aggravate cardiovascular and respiratory diseases, resulting in 23,500 premature deaths annually in the UK [14].
- One ton of hydrogen requires 9 tons of water to produce. In 2016, the production of ammonia reached 146 million tons. Accordingly, 233.6 million tons (233.6 billion liters) of water are needed for ammonia production through water electrolysis, which may cause a "water crisis" in the world [15].

Power-to-ammonia that generates ammonia from renewables can offer an effective way to reduce the intensity of fossil fuels in ammonia synthesis. In this book chapter, we focus on the thermodynamic aspects of renewable ammonia synthesis. For this purpose, we first briefly introduce ammonia synthesis techniques. Second, worldwide demonstrative and simulative projects are presented to give an idea of the thermodynamic and economic view of ammonia synthesis. Then, basic thermodynamic analysis is performed for power-to-ammonia systems. Finally, conclusions and future directions are given in the last section.

11.2 Ammonia Synthesis Pathways

Currently, ammonia has been synthesized through the conventional Haber–Bosch process. In this process, high purity (99.99%) hydrogen reacts with nitrogen in the reactor with catalytic material, e.g., at a temperature between 600 and 830ăK at pressures between 10 and 25 ăMPa. The reaction is highly exothermic:

$$N_2\,(g) + 3H_2\,(g) \rightarrow 2NH_3\,(g)\ \ \Delta H = -92.44\ \mathrm{kJ/mol}. \tag{11.1}$$

In conventional systems, hydrogen comes from hydrocarbons through the steam reforming of methane and the partial oxidation of coal, which makes the process highly carbon intensive. N_2 has typically been separated from air, e.g., by pressure swing adsorption and cryogenic distillation.

Since this method uses fossil fuels like coal and natural gas to make hydrogen, solutions based on renewable energy have been made. Water electrolyzer technologies can be the potential solution for hydrogen sources by using water-splitting from renewable energies, e.g., solar and wind. The water electrolyzer integrated Haber–Bosch system is known as "power-to-ammonia (PtA)," as demonstrated in Figure 11.2.

PtA is widely studied in the literature for the assessment of thermodynamic, techno-economic, and life cycle phenomena. Hasan and Dincer [16] performed a thermodynamic analysis for an integrated wind turbine and solar PV panel with a proton exchange membrane electrolyzer to produce hydrogen for the ammonia synthesis unit. The simulation was run based on the hourly demands of Ontario, Canada, for over 12 months. The results showed that the

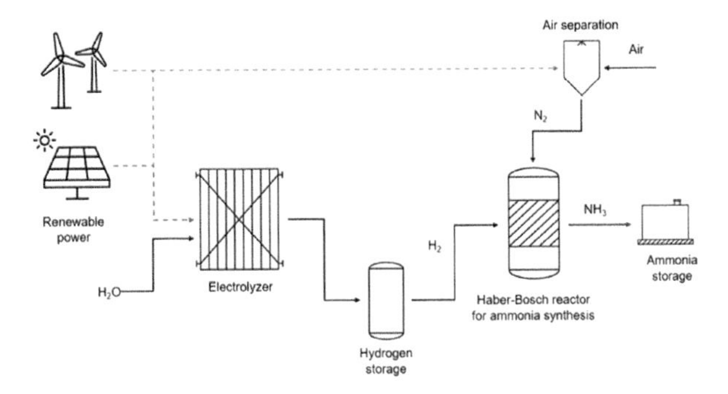

Figure 11.2 A basic configuration of the PtA system process.

system was capable of producing green ammonia (450 kg/h) with an overall system energy and exergy efficiency of 18.8% and 19.1%, respectively. Fasihi et al. [17] focused on the economic performance of the PtA system, which includes a pressurized alkaline electrolyzer (30 bar), a hydrogen storage unit, wind turbines and solar PV panels, and a catalytic reactor. As shown in Figure 11.2, they found that ammonia production costs would decrease significantly for the following decades and become competitive compared to traditional systems due to the projected drop in renewable electricity prices. A small increment in ammonia synthesis cost for conventional systems (coal or natural gas based systems) was also expected because of GHG emissions costs (this effect is not shown in Figure 11.3). Based on the study, the PtA system would be cost-competitive just beyond 2035.

Cheema and Krewer [18] conducted a study to find out how flexible PtA process variables are in terms of how they can be used and how they can be made. Under different loads, six parameters were studied, including the reactor pressure, the amount of inert gas in the synthesis loop, the total flow rate, the amount of NH_3, the temperature of the feed, and the ratio of H_2 to N_2. In the end, the fact that these parameters can be changed means that ammonia production can go down (by up to 73%) or up (by up to 24%). Verleysena et al. [19] studied to answer the following hypothesis question: "How can power-to-ammonia be robust?" For this purpose, they optimized the ammonia synthesis plant powered by wind turbines. Interestingly, their

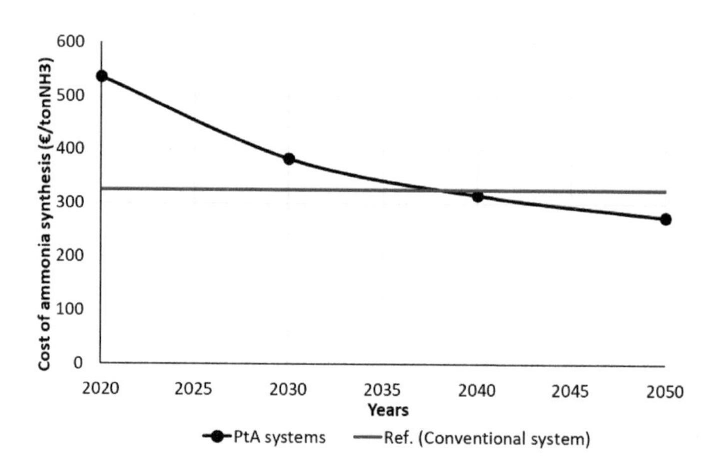

Figure 11.3 On average at the best sites in the world cost of ammonia synthesis through the PtA process (data reproduced based on ref. [17]).

sensitivity analysis showed that the error of wind speed measurement and wind temperature variation (±10 °C) affect ammonia production dramatically. Zhang et al. [13] analyzed the techno-economic performance of solid oxide electrolyzer-based PtA systems. The system achieved a maximum system efficiency of 74%. The payback time was found to be around 5 years, which is still higher than the payback time for conventional systems (2 years). This was because of the high stack costs of electrolyzer and the high electricity prices imported for electrolyzer.

Another ammonia synthesis method is via electrochemical systems, which are either direct N_2 electrolysis or electrochemical N_2 reduction [20], as given in the following equation. This method is an effective approach for ammonia synthesis because it is run under low pressure and temperature (near ambient conditions) using the inexpensive and abundant feedstocks, N_2 and H_2O [21]. However, it is still challenging to be used commercially due to various factors, such as low yields of NH_3, a strong kinetic barrier for N_2 cleavage, high-cost electrolytes, and degradation of electrocatalysts [22, 23].

$$N_2+6H^++6e^-\rightarrow 2NH_3. \tag{11.2}$$

There is a tradeoff for the design of the electrocatalyst. The electrocatalysts for N_2 reduction should have optimal binding energies, with robust binding to N_2 and weak binding to ammonia. Currently, faradic efficiency (which measures between the actual and theoretical maximum amount of ammonia product) of electrochemical ammonia synthesis is lower than 10% owing to very low yields of ammonia. However, some extraordinary studies addressed higher yields and Faradaic efficiency. For instance, Wang et al. [24] found ammonia yield at 7.48 $\mu g\ h^{-1}\ mg^{-1}$ resulting in a faradic efficiency of 55%. Developing strategies for the design of electrocatalysts and electrolytes and optimizing the reaction parameters during the N_2 reduction process would be future goals to make electrochemical ammonia synthesis mechanisms more competitive.

11.3 Worldwide Industry Scale PtA Projects

Table 11.1 gives several worldwide PtA projects that are currently operating, under construction, or being planned. Although numerous PtA projects are very limited so far, many PtA projects will be operated beyond 2023. The PtA, located in North European countries (e.g., Denmark, Germany, and Norway), has gained much interest in the last few decades [25]. The reason is that

Table 11.1 Currently operating, under construction, and planned industry-scale PtA projects.

Industry	**Location**	**Year**	**Power capacity**	**Notes**	**Source** (available date: November 3, 2022)
Ørsted and Yara	Netherlands	2020-	100 MW	75,000 tons of green ammonia per year from Wind energy	https://orsted.com/
Ørsted, Arcelor-Mittal, Yara, Dow Benelux, and Zeeland Refinery	Netherlands and Belgium	N.A.	10 GW	World's largest renewable hydrogen plants to be linked to ammonia production	https://orsted.com/
CF Industries Holdings, Inc.	Louisiana, U.S.	2021-	20 MW	20,000 tons per year of green ammonia	www.cfindustries.com
Topsoe, Vestas, and Skovgaard	Lemvig, Denmark	2023-	62 MW	5000 tons per year	https://blog.topsoe.com/danish-power-to-x-partnership-breaks-ground-on-first-of-its-kind-green-ammonia-project
EDF Renewables and ZeroWaste	Egypt	2026-	N.A.	140,000 tons per year	https://renewablesnow.com
ENGIE and Yara	Australia	2024-	500 MW	800,000 tons per year	https://www.ammoniaenergy.org
Grupo Fertiberia	Puertollano, Spain	2024-	20 MW	200,000 tons per year	https://www.industryandenergy.eu/hyte/hyte-vlog9-europes-largest-in-puertollano/
Jakson Group and Rajasthan state	Kota, India	2028-	N.A.	5,000,000 tons per year	https://timesofindia.indiatimes.com/
H2Carrier and Statkraft	Nordic sea		300 MW	100,000–230,000 tons of green ammonia through a floater	https://www.h2carrier.com/

renewable energy sources (e.g., wind power, hydropower, and biofuels) are abundant due to geographical conditions [25]:

- Wind power accounts for 48% of Danish energy consumption, with a total renewable energy capacity of 10.3 GW reported.
- 92% of the electricity consumption in Norway is provided by hydropower energy, and the total renewable capacity is 39.8 GW.

- Biofuel production is estimated to be 144 TWh in Sweden, and the total renewable energy capacity is 34.6 TWh.

The Danish partnership comprising Topsoe, Skovgaard Energy, and Vestas has launched the PtA project in Lemvig, Denmark. Here, a solid oxide electrolyzer powered by 50 MW of new solar panels and 12 MW of existing V80-2.0 Vestas wind turbines for renewable hydrogen generation will be operational by 2023, producing 5000 tons of green ammonia while saving 8200 tons of CO_2 emissions. The world's first industrial-scale floating green ammonia production facility (the P2XFloater) in the Nordic Sea has been launched by H2Carrier. One of the biggest benefits of having such a system is its flexibility in location, since it can find appropriate places where renewable energy is low-cost. Through the system, it is projected to produce ammonia in the range of 100,000–230,000 tons annually. One of the largest hydrogen plants to be linked to green ammonia production has been built by a large consortium (including Ørsted, Yara, ArcelorMittal, Dow Benelux, Zeeland Refinery, North Sea Port, Smart Delta Resources, and Province of Zeeland and Province of Oost-Vlaanderen) to meet hydrogen and ammonia demand of the Netherlands and Belgium. The electrolyzer, powered by offshore wind capacity with a capacity of 10 GW, is used to generate 1,000,000 tons of renewable hydrogen linked to ammonia synthesis.

11.4 Thermodynamic Assessment of PtA System

Thermodynamic analysis is an effective tool to enhance the performance of energy systems [26]. In thermodynamic analysis, component-level models that include equations originating from the fundamentals of thermodynamics and electrochemistry are developed [27]. Then, communication between component models is provided. The steady-state mass and energy balance applied around the control volume of components is presented by eqn (11.3)–(11.7):

$$\sum \dot{m}_i - \sum \dot{m}_o + \dot{S} = 0 \tag{11.3}$$

$$\sum \dot{m}_i h_i - \sum \dot{m}_o h_o + \dot{Q} = 0 \tag{11.4}$$

where $\dot{S}$ and $\dot{Q}$ represent the mass and energy source/sink terms, respectively. The source/sink terms for each component in the PtA system considered are listed in Table 11.2.

Table 11.2 The sink and source terms used in mass and energy balance for the thermodynamic model.

Components	Sink term	Source term
Electrolyzer	$\dot{S}_{H_2O} = \frac{i}{2F}.A_{cell}.n_{cell}$ [mol/s]	$\dot{S}_{H_2} = \frac{i}{2F}.A_{cell}.n_{cell}$ [mol/s]
	$\dot{S}_{O_2} = \frac{i}{4F}.A_{cell}.n_{cell}$ [mol/s]	$\dot{Q}_{el} = i.V_{cell}.A_{cell}.n_{cell}$ $\left[\frac{W}{m^2}\right]$
Reactor	$\dot{S}_{N_2} = R_{HB}.\dot{V}$	$\dot{S}_{NH_3} = 2.r_{HB}.\dot{V}$
	$\dot{S}_{H_2} = 3.R_{HB}.\dot{V}$	

where i is the current density, A_{cell} is the electrolyzer cell's active area, n_{cell} is the number of cell in the electrolyzer stack, F is Faraday's constant, and R_{HB} is the reaction rate obtained from the Haber–Bosch reactor. There are two approaches to model the reactor such as chemical equilibrium and kinetics. Cheema and Krewer [18] presented the reaction kinetic mechanism for Fe_3O_4 as follows:

$$R_{HB} = k\left(K^2 a_{N_2}\left(\frac{a_{H_2}^3}{a_{NH_3}^2}\right)^a - \left(\frac{a_{NH_3}^2}{a_{H_2}^3}\right)^{1-a}\right) \tag{11.5}$$

where a_{N_2}, a_{H_2}, and a_{NH_3} represent activity coefficients for nitrogen, hydrogen, and ammonia, respectively. k, K, and a are the constant for the reverse reaction, equilibrium constant, and constant, respectively. k can be calculated by using the Arrhenius equation as given in the following equation:

$$k = k_0.e^{-E_a/RT} \tag{11.6}$$

where k_0 represents the preexponential factor and E_a is the activation energy for ammonia decomposition. The value of parameters is given in Table 11.3.

Alternatively, chemical equilibrium is an effective approach to calculate the outlet composition of the reactor by incorporating mass balance. This approach is based on Le Chatelier's principle that determines the reaction direction. For instance, the reaction goes to the right (product side) when N_2 or H_2 is added, while the reaction shifts to the reactant side when NH_3 is added.

Table 11.3 Reaction kinetics parameters [18].

Parameter	**Value**
k_0 $\left[\frac{kmol}{m^3}\right]$	$8.8490 \cdot 10^{14}$
E_a $\left[\frac{kJ}{kmol}\right]$	$1.7056 \cdot 10^5$
a	0.5

After mass and energy balance are applied around the control volume of the components, the model is ready to estimate electrical efficiency as follows:

$$\eta_{\text{eff}} = \frac{\dot{n}_{\text{NH}_3}.\text{LHV}}{P_{\text{el}}} \tag{11.7}$$

where LHV is the lower heating value of ammonia and P_{el} is the required power for the electrolyzer.

In this chapter, we study the thermodynamic analysis of a 1-MW PtA system electrolyzer under various working temperatures and pressures. The system's hydrogen requirement is provided by a water electrolyzer. High- and low-temperature electrolyzers are separately implemented into the system to understand the electrolyzer's impact on the PtA system's performance. However, it should be noted that the comparison is performed considering electrochemical performance, while thermal performance is not accounted for.

The modeling framework is methodologically presented in Figure 11.4. Here, the model of electrolyzers is developed based on the available polarization curve obtained from Graves et al. [28]. The thermoneutral voltage (E_{th}) in the water electrolysis process represents the irreversible heat of evaporation for the phase change of liquid water to hydrogen and oxygen, which are gaseous forms, a contribution that is not considered in the reversible voltage. Water electrolysis becomes an endothermic process when cell voltages are less than the thermoneutral voltage, while for cell voltages bigger than the thermoneutral voltage, water electrolysis is an exothermic process [29]. In this regard, we select the thermoneutral voltage as an operating voltage that provides the operation without heat loss to the surroundings or heat adsorption from the surroundings. This voltage and power capacity determine the size of the electrolyzer. The flow rates of hydrogen and water that are produced and consumed, respectively, during the electrolysis process are calculated using Faraday's law. The amount of nitrogen fed to a chemical Haber–Bosch reactor is estimated based on the amount of hydrogen. The reactor is modeled under the chemical equilibrium of the Haber–Bosch reaction. The mass balance is then applied around the control volume of the reactor to calculate the amount of hydrogen, nitrogen, and ammonia used during the process. The model is developed in the Cantera MATLAB environment. The results are presented below.

The reactor is modeled using chemical equilibrium. In Figure 11.5, the change in reactor temperature regarding reactor inlet temperature under

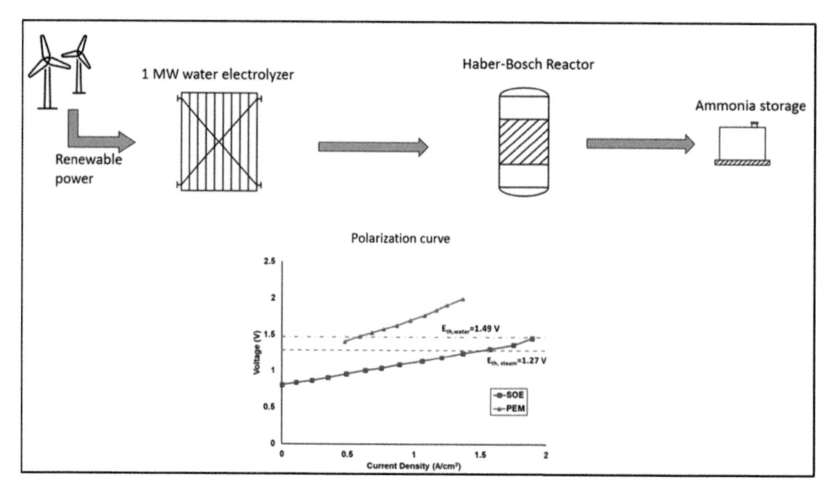

Figure 11.4 A schematic configuration of the method considered in this study.

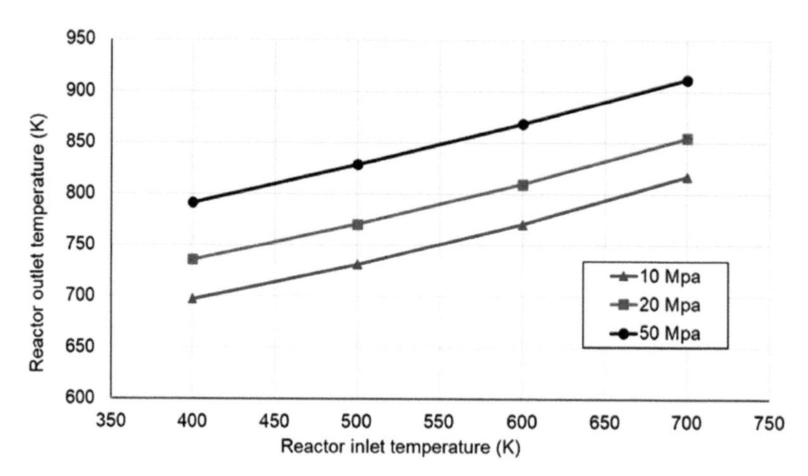

Figure 11.5 The change of reactor outlet temperature versus reactor inlet temperature under various operating pressures.

different operating pressures is given. The pressure boosts the reactor outlet temperature significantly. For example, when the operating pressure increases from 10 to 50 MPa, the increase in reactor outlet temperature may be up to 100 K. Moreover, the maximum reactor outlet is found as 900 K when the reactor inlet temperature and pressure are 700 K and 50 MPa, respectively.

In Figure 11.6, the reactor outlet molar compositions for hydrogen, nitrogen, and ammonia are given for various reactor inlet temperatures

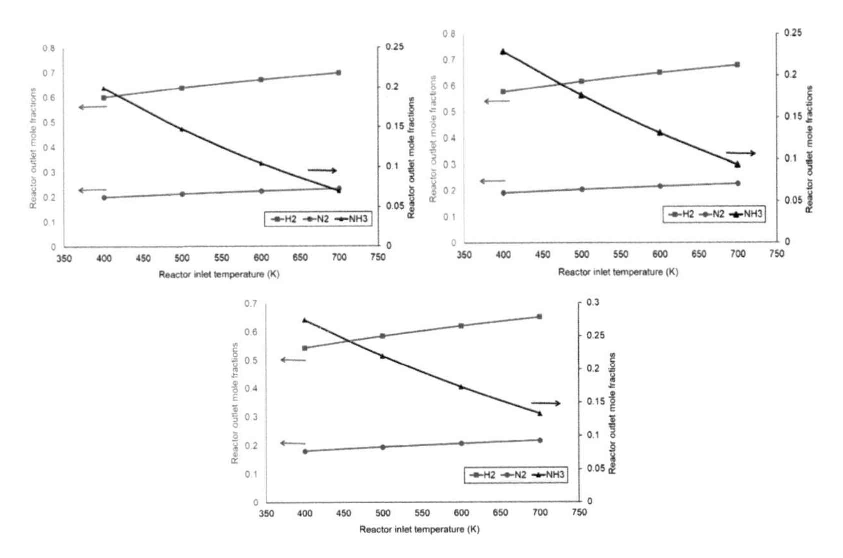

Figure 11.6 The effect of pressure on the reactor outlet compositions: (a) 10 MPa, (b) 20 MPa, and (c) 50 MPa.

and pressures. As the pressure increases, the chemical activity increases, promoting the ammonia yield. For instance, the ammonia mole fraction increases from 0.17 to 0.27 when the pressure increases from 10 to 50 MPa, while hydrogen utilization increases by 20% at a reactor inlet temperature of 400 K. However, the minimum mole fraction of ammonia is calculated as 0.06 at the reactor's inlet temperature and pressure of 700 K and 10 MPa, respectively. Therefore, low reactor inlet temperature and high operating pressure promote the reaction activity that increases ammonia yield. Meanwhile, there has been no noticeable change in nitrogen utilization. The reason may be because the N_2 cleavage is very difficult to separate.

Figure 11.7 shows how the type of electrolyzer affects the flow rates of ammonia made and nitrogen used when the electrolyzers are run at their thermoneutral voltages for different operating pressures. When compared to PEM electrolyzer-integrated PtA systems, SOE systems are better because they can make more ammonia. Moreover, higher nitrogen utilization can be obtained when SOE is used. It is because the amount of hydrogen produced decreases under PEM electrolyzer operation (at their thermoneutral voltages), which results in a lower amount of nitrogen fed to the reactor. Even though the nitrogen mole fraction at the reactor outlet is not significantly changed (as we confirmed in Figure 11.6), as the inlet amount of nitrogen decreases,

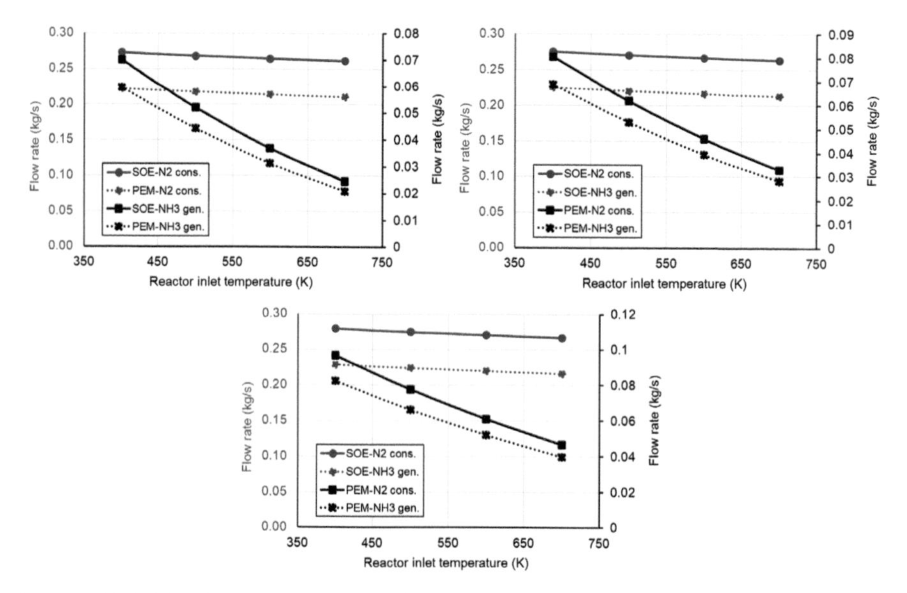

Figure 11.7 The flow rates of ammonia synthesized and nitrogen consumed during the reaction under the operation of SOE and PEM electrolyzer for pressures of (a) 10 MPa, (b) 20 MPa, and (c) 50 MPa.

the nitrogen utilization decreases. However, particularly, this effect decreases when the reactor inlet temperature increases (e.g., 700 K). It should be noted that the temperature range depends on the electrocatalyst selected. Moreover, water consumption is also influenced by the selection of the electrolyzer type. In this regard, we found that water consumption can increase by 20% for the SOE-integrated PtA system compared to PEM-based PtA systems.

11.5 Conclusions and Future Directions

In this chapter, we present an effective way of utilizing hydrogen by producing green ammonia. After the opportunities and drawbacks in front of the ammonia industry are briefly discussed, the ammonia synthesis pathways are introduced, such as conventional systems, power-to-ammonia, and electrochemical ammonia synthesis.

As a point of economy, the findings taken from the literature show that the power-to-ammonia option can become cost-competitive in the near future (particularly beyond 2030), which highly depend on the renewable electricity generation cost and electrolyzer stack costs. We also show the feasibility and capability of industry-scale power-to-ammonia projects that are currently operating, under construction, or being planned around the world. Although the number of commercial projects is very limited before 2020, there is a significant interest in constructing and getting power-to-ammonia projects operational in the near future. Especially in the northern countries of Europe, considerable attention has been given to PtA systems because a significant increase in renewable energy in those countries leads to the need for electricity storage. For example, the PtA system, which includes a solid oxide electrolyzer powered by hybrid wind turbines and solar PV panels with a capacity of 62 MW will be operational by 2023 to produce 5000-ton green ammonia in Denmark. This system is planning to contribute to saving 8200 tons of CO_2 emissions from being released into the atmosphere. To gain flexibility in its location, there is also a floating PtA ship. Hence, the system can be used in appropriate places where renewable energy is of low cost. We finally performed a basic thermodynamic analysis for the coupled water electrolyzer and Haber–Bosch system under various working conditions. Here, the effect of the electrolyzer type used (SOE or PEM) on the PtA system performance is considered. Based on this analysis, the main remarkable results can be taken as follows:

- The reactor outlet temperature may increase up to 900 K when the reactor inlet temperature and pressure are 700 K and 50 MPa, respectively, during the Haber–Bosch operation.
- Low reactor inlet temperature and high operating pressure promote the reaction activity that increases ammonia yield.
- The use of SOE in the system has the capability of producing more amount of ammonia than PEM electrolyzer-based systems.
- Water consumption increases by 20% for SOE-integrated PtA systems compared to PEM electrolyzer-based systems.

This study confirms that PtA will gain more traction in the near future. However, the PtA process is still lacking in the use of valuable feedstocks such as hydrogen. Moreover, the system produces ammonia with a few steps like conventional systems (e.g., renewable energy to hydrogen and hydrogen to ammonia), which results in low system overall efficiency, high cost, and need for a lot of space. Alternatively, electrochemical synthesis with the reduction of nitrogen can be a promising technology for green ammonia synthesis at atmospheric temperature and pressure using inexpensive and abundant feedstocks such as nitrogen and water.

References

[1] M. Hulme, 1.5 °C and climate research after the Paris Agreement, Nat. Clim. Change. 6 (2016) 222–224. https://doi.org/10.1038/nclimate2939.

[2] L. Kemp, C. Xu, J. Depledge, K.L. Ebi, G. Gibbins, T.A. Kohler, J. Rockström, M. Scheffer, H.J. Schellnhuber, W. Steffen, T.M. Lenton, Climate Endgame: Exploring catastrophic climate change scenarios, Proc. Natl. Acad. Sci. 119 (2022) e2108146119. https://doi.org/10.1073/pnas.2108146119.

[3] M. Meinshausen, J. Lewis, C. McGlade, J. Gütschow, Z. Nicholls, R. Burdon, L. Cozzi, B. Hackmann, Realization of Paris Agreement pledges may limit warming just below 2 °C, Nature. 604 (2022) 304–309. https://doi.org/10.1038/s41586-022-04553-z.

[4] World Energy Outlook 2020 – Analysis - IEA, (n.d.). https://www.iea.org/reports/world-energy-outlook-2020 (accessed March 25, 2021).

[5] IEA – International Energy Agency, IEA. (n.d.). https://www.iea.org (accessed July 31, 2021).

[6] M.M. Rahman, A.O. Oni, E. Gemechu, A. Kumar, Assessment of energy storage technologies: A review, Energy Convers. Manag. 223 (2020) 113295. https://doi.org/10.1016/j.enconman.2020.113295.

[7] A.K. Shukla, O. Singh, M. Sharma, R.K. Phanden, J.P. Davim, eds., Hybrid power cycle arrangements for lower emissions, CRC Press/Balkema, Leiden, The Netherlands, 2022.

[8] J.S. Cardoso, V. Silva, R.C. Rocha, M.J. Hall, M. Costa, D. Eusébio, Ammonia as an energy vector: Current and future prospects for low-carbon fuel applications in internal combustion engines, J. Clean. Prod. 296 (2021) 126562. https://doi.org/10.1016/j.jclepro.2021.126562.

[9] A.C. Ince, C.O. Colpan, A. Hagen, M.F. Serincan, Modeling and simulation of Power-to-X systems: A review, Fuel. 304 (2021) 121354. https://doi.org/10.1016/j.fuel.2021.121354.

[10] A.C. Ince, C. Ozgur Colpan, A. Keles, M.F. Serincan, U. Pasaogullari, Scaling and performance assessment of power-to-methane system based on an operation scenario, Fuel. 332 (2023) 126182. https://doi.org/10.1016/j.fuel.2022.126182.

[11] P.M. Falcone, M. Hiete, A. Sapio, Hydrogen economy and sustainable development goals: Review and policy insights, Curr. Opin. Green Sustain. Chem. 31 (2021) 100506. https://doi.org/10.1016/j.cogsc.2021.100506.

[12] J.W. Makepeace, T. He, C. Weidenthaler, T.R. Jensen, F. Chang, T. Vegge, P. Ngene, Y. Kojima, P.E. de Jongh, P. Chen, W.I.F. David, Reversible ammonia-based and liquid organic hydrogen carriers for high-density hydrogen storage: Recent progress, Int. J. Hydrog. Energy. 44 (2019) 7746–7767. https://doi.org/10.1016/j.ijhydene.2019.01.144.

[13] H. Zhang, L. Wang, J. Van herle, F. Maréchal, U. Desideri, Techno-economic comparison of green ammonia production processes, Appl. Energy. 259 (2020) 114135. https://doi.org/10.1016/j.apenergy.2019.114135.

[14] A. Valera-Medina, H. Xiao, M. Owen-Jones, W.I.F. David, P.J. Bowen, Ammonia for power, Prog. Energy Combust. Sci. 69 (2018) 63–102. https://doi.org/10.1016/j.pecs.2018.07.001.

[15] S. Ghavam, M. Vahdati, I.A.G. Wilson, P. Styring, Sustainable Ammonia Production Processes, Front. Energy Res. 9 (2021) 580808. https://doi.org/10.3389/fenrg.2021.580808.

[16] A. Hasan, I. Dincer, Development of an integrated wind and PV system for ammonia and power production for a sustainable community, J. Clean. Prod. 231 (2019) 1515–1525. https://doi.org/10.1016/j.jclepro.2019.05.110.

[17] M. Fasihi, R. Weiss, J. Savolainen, C. Breyer, Global potential of green ammonia based on hybrid PV-wind power plants, Appl. Energy. 294 (2021) 116170. https://doi.org/10.1016/j.apenergy.2020.116170.

[18] I.I. Cheema, U. Krewer, Operating envelope of Haber–Bosch process design for power-to-ammonia, RSC Adv. 8 (2018) 34926–34936. https://doi.org/10.1039/C8RA06821F.

[19] K. Verleysen, D. Coppitters, A. Parente, W. De Paepe, F. Contino, How can power-to-ammonia be robust? Optimization of an ammonia synthesis plant powered by a wind turbine considering operational uncertainties, Fuel. 266 (2020) 117049. https://doi.org/10.1016/j.fuel.2020.117049.

[20] F. Jiao, B. Xu, Electrochemical Ammonia Synthesis and Ammonia Fuel Cells, Adv. Mater. 31 (2019) 1805173. https://doi.org/10.1002/adma.201805173.

[21] R. Zhao, H. Xie, L. Chang, X. Zhang, X. Zhu, X. Tong, T. Wang, Y. Luo, P. Wei, Z. Wang, X. Sun, Recent progress in the electrochemical ammonia synthesis under ambient conditions, EnergyChem. 1 (2019) 100011. https://doi.org/10.1016/j.enchem.2019.100011.

[22] Y. Lu, J. Li, T. Tada, Y. Toda, S. Ueda, T. Yokoyama, M. Kitano, H. Hosono, Water Durable Electride Y 5 Si 3ă: Electronic Structure and

Catalytic Activity for Ammonia Synthesis, J. Am. Chem. Soc. 138 (2016) 3970–3973. https://doi.org/10.1021/jacs.6b00124.

[23] H. Shen, C. Choi, J. Masa, X. Li, J. Qiu, Y. Jung, Z. Sun, Electrochemical ammonia synthesis: Mechanistic understanding and catalyst design, Chem. 7 (2021) 1708–1754. https://doi.org/10.1016/j.chempr.2021.01.009.

[24] M. Wang, S. Liu, T. Qian, J. Liu, J. Zhou, H. Ji, J. Xiong, J. Zhong, C. Yan, Over 56.55% Faradaic efficiency of ambient ammonia synthesis enabled by positively shifting the reaction potential, Nat. Commun. 10 (2019) 341. https://doi.org/10.1038/s41467-018-08120-x.

[25] J. Ikäheimo, J. Kiviluoma, R. Weiss, H. Holttinen, Power-to-ammonia in future North European 100 % renewable power and heat system, Int. J. Hydrog. Energy. 43 (2018) 17295–17308. https://doi.org/10.1016/j.ijhydene.2018.06.121.

[26] I. Dincer, Thermodynamics: A Smart Approach, John Wiley & Sons, 2020.

[27] A.C. Ince, M.U. Karaoglan, A. Glüsen, C.O. Colpan, M. Müller, D. Stolten, Semiempirical thermodynamic modeling of a direct methanol fuel cell system, Int. J. Energy Res. 43 (2019) 3601–3615. https://doi.org/10.1002/er.4508.

[28] C. Graves, S.D. Ebbesen, M. Mogensen, K.S. Lackner, Sustainable hydrocarbon fuels by recycling CO 2 and H 2 O with renewable or nuclear energy, Renew. Sustain. Energy Rev. 15 (2011) 1–23. https://doi.org/10.1016/j.rser.2010.07.014.

[29] C. Lamy, P. Millet, A critical review on the definitions used to calculate the energy efficiency coefficients of water electrolysis cells working under near ambient temperature conditions, J. Power Sources. 447 (2020) 227350. https://doi.org/10.1016/j.jpowsour.2019.227350.

Biographies

Alper Can Ince received the bachelor's degree in mechanical engineering the master's degree in mechanical engineering from Dokuz Eylul University in 2017 and 2019, respectively. He worked as a Research Assistant with the Department of Mechanical Engineering, Gebze Technical University between 2018 and 2022. He has begun his Ph.D. studies at the Department of Mechanical Engineering, Faculty of Engineering, University of Connecticut in 2022. His research areas include fuel cells, hydrogen, and energy conversion and storage systems. He has been serving as a Reviewer for many highly respected journals.

Can Özgür Çolpan received his bachelor's and master's degrees from the Mechanical Engineering Department of the Middle East Technical University in 2003 and 2005, respectively. He conducted his Ph.D. studies at the Department of Mechanical and Aerospace Engineering of Carleton University, in Ottawa, Canada between 2005 and 2009. He continued his studies as a postdoctoral researcher at the same department between 2009 and 2010 and at the Mechanical and Industrial Engineering Department of Ryerson University, in Toronto, Canada between 2010 and 2012. He has been working at the Department of Mechanical Engineering of Dokuz Eylül University since 2012. In 2014 and 2019, he was appointed to the positions of Associate Professor and Professor, respectively. Prof. Can Özgür Çolpan has conducted research in the field of fuel cells and hydrogen and the mathematical modeling of integrated energy systems.

Mustafa Fazıl Serincan received the bachelor's degree in mechanical engineering from Istanbul Technical University in 2003, the master's degree in electrical and electronics engineering and computer science from Sabanci University in 2005, and the philosophy of doctorate degree in mechanical engineering from University of Connecticut in 2009. He is currently working as an Associate Professor with the Department of Mechanical Engineering, Faculty of Engineering, Gebze Technical University. His research areas include electrochemical energy conversion, heat exchangers, electronics cooling, and engineering solutions for transport phenomena problems with multiphysics CFD modeling.

Ugur Pasaogullari received his Ph.D. degree in mechanical engineering from the Pennsylvania State University in 2005. He earned his B.S. degree in mechanical engineering with high honors from Middle East Technical University, Ankara, Turkey and the M.S. degree in mechanical engineering from the Pennsylvania State University in 1999 and 2003, respectively. His current research focuses on thermal and water management in polymer electrolyte fuel cells, and characterization of the effects of heat and water transport on performance and durability of these systems. He was appointed as Castleman term professor in 2016.

Index

C
Cyclohexanes 147
Combined cycle 208

E
Energy 3, 60, 266
Engine 96, 163, 171
Emission 3, 99, 176

F
Fuel cells 18, 93, 238
Fuel cell electric vehicles 81, 98
Formic acid 94, 146, 151

G
Green hydrogen 5, 131, 140
Gas turbine 13, 187, 200
GT 184, 201, 213
Green ammonia 9, 250, 257

H
Hydrogen technologies 85, 239
Heterogeneous catalysts 146, 153
Homogeneous catalysts 146, 152
HCNG 163, 173, 178
Hydrogen 12, 82, 91
Hydrogen fuel 9, 186, 211

L
Liquid organic hydrogen carriers 92, 146

M
Membrane 17, 98, 247
Modeling 57, 213

M
N-heterocycles 156

O
Offshore wind energy 123, 134, 141

P
Polymer 17, 23, 34
Proton conductivity 17, 29, 36
PEMFC 17, 30, 55
PEM fuel cell 19, 57, 211
PESA-II 55, 64, 75
Prospects 81, 108
Performance 23, 36, 55
Power plant 9, 127, 133
Power generation 1, 133, 185
Power-to-ammonia 243, 248, 257

R
Regenerative ORC 53, 59, 75
Roadmaps 8, 82, 108
Refueling 82, 96, 107

S

Ships 103, 123, 138
Synthetic fuels 86, 123, 137
SOFC 93, 104, 207

T

Tri-objective optimization 53, 63, 70
TOPSIS 53, 69, 75
Thermodynamic analysis 156, 247, 257

W

Water electrolyzer 211, 247, 253

About the Editors

Dr. Anoop Kumar Shukla is presently working as an assistant professor in the Department of Mechanical Engineering at Amity University Noida, India. He completed his M.Tech. with distinction from Harcourt Butler Technological Institute (HBTI) Kanpur. Dr. Shukla earned his Ph.D. from Dr. APJ Abdul Kalam Technical University Lucknow in 2017. His area of research is energy conversion and thermal management, combined power cycles, cogeneration, multigeneration, biofuels, and gas turbine cooling. He has published two edited books and more than 54 SCI/Scopus indexed journal and conference publications. He has guided one Ph.D. and more than twenty projects/dissertations at the UG/PG level. Currently, he is guiding three Ph.D. and two M.Tech. students. He has also authored book chapters with highly acclaimed publishers, including Springer and Material Research Forum. Dr. Shukla has organized several international conferences and symposiums. He is an active reviewer for various highly reputed international journals, including Applied Thermal Engineering, Sustainable Energy Technologies, and Assessments, Applied Sciences, Energy, International Journal of Green Energy, Sustainability, Desalination, Journal of Power and Energy, Thermal Science and Engineering Progress, and Energies.

Prof. Onkar Singh, currently the Vice Chancellor of Veer Madho Singh Bhandari Uttarakhand Technical University, Dehradun is a Professor of Mechanical Engineering at Harcourt Butler Technical University (Formerly HBTI), Kanpur since January 2007. Dr. Singh was also founding Vice Chancellor of Madan Mohan Malaviya University of Technology, Gorakhpur and Vice Chancellor of Uttar Pradesh Technical University, Lucknow. Dr. Singh has authored 11 books and 11 chapters in different books. He has published 231 papers in international/national journals and conferences and 117 popular articles. Dr. Singh has guided 14 Ph.D. and 27 PG dissertations, held 3 patents, and completed 7 research projects. He holds the National Records-LIMCA Book of Records, March 2014 & LIMCA Book of Records, February 2015. Dr. Singh is Fellow and member of number of international and national professional societies of repute.

Prof. Ali J. Chamkha, is currently associated as the Dean of Engineering and in charge of all academic and research affairs at Kuwait College of Science and Technology. He has extensive research experience and authored and co-authored more than 750 peer-reviewed journal and conference publications. He is editor in chief for the Journal of Nanofluids. He also serves as Associate Editor for several journals of high repute, such as ASME Journal of Thermal Science and Engineering Applications, International Journal of Numerical Methods for Heat and Fluid Flow, Journal of Porous Media, Scientia Iranica. Professor Chamkha is the Trainer for Engineering and Management Courses such as leadership skills, soft skills, six sigma, internal auditing, ISO certification, quality control, strategic planning, project and operations management, total quality management, effective presentations and time management skills, seven habits of highly effective people, valves and pumps technology, filtration technology and others. He has received more than 12 achievement awards for outstanding teaching and research from the local, regional and international levels. Professor Chamkha also provides consultancy to various industries such as filtration, water, oil and gas, HSE, environment, quality management, and auditing.

Dr. Meeta Sharma, is working as an Associate Professor in the Mechanical Engineering Department, Amity University, Noida. She completed her B.Tech. and M.Tech. from Aligarh Muslim University, Aligarh, India and her Ph.D. from Dr. A.P.J Abdul Kalam Technical University, Lucknow, India. She has around 18 years of teaching experience and has published good research papers in reputed journals such as Applied Thermal Engineering, Heat Transfer – Asian Research, Journal of Thermal Engineering, Energy Sources and many more. Dr. Sharma has reviewed many research papers for reputed journals such as Applied Thermal Engineering, Heat Transfer – Asian Research, Journal of Thermal Engineering, Energy Sources, Energy Report and Journal of Power and Energy, Part A of the Proceedings of the Institution of Mechanical Engineers. She has contributed to many book chapters and participated in international Scopus conferences. Her areas of interest include combined cycle power plants, cogeneration plants, waste heat recovery systems, solar energy, heat transfer and applied thermodynamics. Currently, Dr. Sharma is guiding two Ph.D. and 3 M.Tech. dissertations and one B.Tech. project as research assignments. She has guided 12 M.Tech. dissertations and many B.Tech. projects. She is a member of the Institution of Engineers.

For Product Safety Concerns and Information please contact our
EU representative GPSR@taylorandfrancis.com Taylor & Francis
Verlag GmbH, Kaufingerstraße 24, 80331 München, Germany